# User's Manual for the National Water-Quality Assessment Program Invertebrate Data Analysis System (IDAS) Software: Version 5

By Thomas F. Cuffney and Robin A. Brightbill

National Water-Quality Assessment Program

Techniques and Methods 7–C4

**U.S. Department of the Interior**
**U.S. Geological Survey**

**U.S. Department of the Interior**
KEN SALAZAR, Secretary

**U.S. Geological Survey**
Marcia K. McNutt, Director

U.S. Geological Survey, Reston, Virginia: 2011

For more information on the USGS—the Federal source for science about the Earth, its natural and living resources, natural hazards, and the environment, visit *http://www.usgs.gov* or call 1-888-ASK-USGS

For an overview of USGS information products, including maps, imagery, and publications, visit *http://www.usgs.gov/pubprod*

To order this and other USGS information products, visit *http://store.usgs.gov*

Suggested citation:
Cuffney, T.F., and Brightbill, R.A., 2011, User's manual for the National Water-Quality Assessment Program Invertebrate Data Analysis System (IDAS) software, version 5: U.S. Geological Survey Techniques and Methods 7–C4, 126 p. (available only online at *http://pubs.usgs.gov/tm/7c4/*)

# Foreword

The U.S. Geological Survey (USGS) is committed to providing the Nation with reliable scientific information that helps to enhance and protect the overall quality of life and that facilitates effective management of water, biological, energy, and mineral resources (*http://www.usgs.gov/*). Information on the Nation's water resources is critical to ensuring long-term availability of water that is safe for drinking and recreation and is suitable for industry, irrigation, and fish and wildlife. Population growth and increasing demands for water make the availability of that water, measured in terms of quantity and quality, even more essential to the long-term sustainability of our communities and ecosystems.

The USGS implemented the National Water-Quality Assessment (NAWQA) Program in 1991 to support national, regional, State, and local information needs and decisions related to water-quality management and policy (*http://water.usgs.gov/nawqa/*). The NAWQA Program is designed to answer: What is the quality of our Nation's streams and groundwater? How are conditions changing over time? How do natural features and human activities affect the quality of streams and groundwater, and where are those effects most pronounced? By combining information on water chemistry, physical characteristics, stream habitat, and aquatic life, the NAWQA Program aims to provide science-based insights for current and emerging water issues and priorities. From 1991 to 2001, the NAWQA Program completed interdisciplinary assessments and established a baseline understanding of water-quality conditions in 51 of the Nation's river basins and aquifers, referred to as Study Units (*http://water.usgs.gov/nawqa/studyu.html*).

National and regional assessments are ongoing in the second decade (2001–2012) of the NAWQA Program as 42 of the 51 Study Units are selectively reassessed. These assessments extend the findings in the Study Units by determining water-quality status and trends at sites that have been consistently monitored for more than a decade, and filling critical gaps in characterizing the quality of surface water and groundwater. For example, increased emphasis has been placed on assessing the quality of source water and finished water associated with many of the Nation's largest community water systems. During the second decade, NAWQA is addressing five national priority topics that build an understanding of how natural features and human activities affect water quality, and establish links between *sources* of contaminants, the *transport* of those contaminants through the hydrologic system, and the potential *effects* of contaminants on humans and aquatic ecosystems. Included are studies on the fate of agricultural chemicals, effects of urbanization on stream ecosystems, bioaccumulation of mercury in stream ecosystems, effects of nutrient enrichment on aquatic ecosystems, and transport of contaminants to public-supply wells. In addition, national syntheses of information on pesticides, volatile organic compounds (VOCs), nutrients, trace elements, and aquatic ecology are continuing.

The USGS aims to disseminate credible, timely, and relevant science information to address practical and effective water-resource management and strategies that protect and restore water quality. We hope this NAWQA publication will provide you with insights and information to meet your needs, and will foster increased citizen awareness and involvement in the protection and restoration of our Nation's waters.

The USGS recognizes that a national assessment by a single program cannot address all water-resource issues of interest. External coordination at all levels is critical for cost-effective management, regulation, and conservation of our Nation's water resources. The NAWQA Program, therefore, depends on advice and information from other agencies—Federal, State, regional, interstate, Tribal, and local—as well as nongovernmental organizations, industry, academia, and other stakeholder groups. Your assistance and suggestions are greatly appreciated.

William H. Werkheiser
Associate Director for Water

# Preface

This report presents a computer program for processing invertebrate samples collected as part of the U.S. Geological Survey (USGS) National Water-Quality Assessment (NAWQA) Program. The performance of this program has been tested using data collected by the NAWQA Program. However, future applications of the program could reveal errors that were not detected in the test simulations. Users are requested to notify the USGS if errors are found in the report or in the computer program. Correspondence regarding the report should be sent to:

> U.S. Geological Survey
> North Carolina Water Science Center
> 3916 Sunset Ridge Road
> Raleigh, NC 27607

Although this program has been used by the USGS, no warranty, expressed or implied, is made by the USGS or the United States Government as to the accuracy and functioning of the program and the related program material. Nor shall the fact of distribution constitute any such warranty, and no responsibility is assumed by the USGS in connection therewith.

The computer program documented in this report is available online at:

*ftp://ftpext.usgs.gov/pub/er/nc/raleigh/tfc/IDAS_v5/*

# Contents

# Figures

## Tables

# Conversion Factors

| Multiply | By | To obtain |
|---|---|---|
| Length | | |
| centimeter (cm) | 0.3937 | inch (in.) |
| meter (m) | 3.281 | foot (ft.) |
| Area | | |
| square centimeter (cm$^2$) | 0.155 | square inch (in$^2$) |
| square meter (m$^2$) | 10.76 | square foot (ft$^2$) |

# Abbreviations frequently used in this report

| | |
|---|---|
| A | adult lifestage |
| ALBE | Albemarle-Pamlico Drainage Basin |
| ALMN | Allegheny and Monongahela Drainage Basin |
| Ambig | ambiguous taxon |
| BG | Biological Group of the NWQL |
| Bio-TDB | Biological Transactional Database |
| BU_ID | taxon name applied by BG |
| cf. | confer |
| dam. | damaged |
| DPAC | distribute parent among children |
| DTH | depositional targeted habitat |
| IDAS | Invertebrate Data Analysis System |
| imm. | immature |
| indet. | indeterminate |
| KB | kilobyte |
| L | larval lifestage |
| MB | megabyte |
| MCWP | merge child with parent |
| MHz | megahertz |
| NAWQA | National Water-Quality Assessment |
| NCDENR | North Carolina Department of Environment and Natural Resources |
| no. | number |
| NWQL | National Water Quality Laboratory |
| ORIG | original |
| P | pupal lifestage |
| QMH | qualitative multihabitat |
| QUAL | qualitative sample: RTH+DTH+QMH |
| RAM | random access memory |
| RBP | Rapid Bioassessment Protocol |
| ref. | reference |
| RPKC | remove parent, keep child |
| RPMC | remove parent or merge child |
| RTH | Richest targeted habitat |
| SampleID | sample identifier used by Bio-TDB |
| SMCOD | sample identification code |
| SortCode | taxonomic sort code |
| sp. | species |
| STAID | station identifier |
| SUID | 4-character Study Unit identifier |
| TOL | tolerance values |
| USEPA | U.S. Environmental Protection Agency |
| USGS | U.S. Geological Survey |
| WR-EMAP | Western Region Environmental Monitoring and Assessment Program |
| % | percentage |
| + | plus |
| ± | plus or minus |
| = | equal to |
| nr. | near |
| < | less than |
| > | greater than |
| ≤ | less than or equal to |
| ≥ | greater than or equal to |

# User's Manual for the National Water-Quality Assessment Program Invertebrate Data Analysis System (IDAS) Software: Version 5

By Thomas F. Cuffney and Robin A. Brightbill

## Abstract

The Invertebrate Data Analysis System (IDAS) software was developed to provide an accurate, consistent, and efficient mechanism for analyzing invertebrate data collected as part of the U.S. Geological Survey National Water-Quality Assessment (NAWQA) Program. The IDAS software is a stand-alone program for personal computers that run Microsoft Windows®. It allows users to read data downloaded from the NAWQA Program Biological Transactional Database (Bio-TDB) or to import data from other sources either as Microsoft Excel® or Microsoft Access® files. The program consists of five modules: Edit Data, Data Preparation, Calculate Community Metrics, Calculate Diversities and Similarities, and Data Export. The Edit Data module allows the user to subset data on the basis of taxonomy or sample type, extract a random subsample of data, combine or delete data, summarize distributions, resolve ambiguous taxa (see glossary) and conditional/provisional taxa, import non-NAWQA data, and maintain and create files of invertebrate attributes that are used in the calculation of invertebrate metrics. The Data Preparation module allows the user to select the type(s) of sample(s) to process, calculate densities, delete taxa on the basis of laboratory processing notes, delete pupae or terrestrial adults, combine lifestages or keep them separate, select a lowest taxonomic level for analysis, delete rare taxa on the basis of the number of sites where a taxon occurs and (or) the abundance of a taxon in a sample, and resolve taxonomic ambiguities by one of four methods. The Calculate Community Metrics module allows the user to calculate 184 community metrics, including metrics based on organism tolerances, functional feeding groups, and behavior. The Calculate Diversities and Similarities module allows the user to calculate nine diversity and eight similarity indices. The Data Export module allows the user to export data to other software packages (CANOCO, Primer, PC-ORD, MVSP) and produce tables of community data that can be imported into spreadsheet, database, graphics, statistics, and word-processing programs. The IDAS program facilitates the documentation of analyses by keeping a log of the data that are processed, the files that are generated, and the program settings used to process the data. Though the IDAS program was developed to process NAWQA Program invertebrate data downloaded from Bio-TDB, the Edit Data module includes tools that can be used to convert non-NAWQA data into Bio-TDB format. Consequently, the data manipulation, analysis, and export procedures provided by the IDAS program can be used to process data generated outside of the NAWQA Program.

## Introduction

The U.S. Geological Survey's (USGS) National Water-Quality Assessment (NAWQA) Program is a perennial program designed to provide a comprehensive, interdisciplinary water-quality assessment of the Nation's flowing water resources (Hirsch and others, 1988; Leahy and others, 1990). During the first decade of the Program's operation (1991–2001), ecological studies were conducted to assess the occurrence and distribution of algal, invertebrate, and fish communities in about 51 Study Units (Gilliom and others, 1995). These Study Units represent the dominant hydrologic systems nationwide and are staggered in time with respect to implementation and high- and low-intensity sampling periods (Gilliom and others, 1995). During the second decade of the NAWQA Program (2001–2011), ecological studies are being conducted as part of nationally guided topical studies that address selected water-quality issues and as part of long-term-trends monitoring within the Study Units.

Accurate, consistent, and timely analyses of invertebrate data are needed to support water-quality investigations within the NAWQA Program. A common set of data-analysis tools and taxa attributes makes analyses easier, more consistent, and more accurate while providing better documentation and archiving of results and methods of analysis. These tools can also facilitate communication and interactions among scientific teams that are geographically dispersed (for example, study unit, regional, and national teams) both within and outside of the USGS. Data analysis tools, like the Invertebrate Data Analysis System (IDAS), that can utilize data in a variety of formats make it easier for the NAWQA Program to incorporate non-NAWQA data in its analyses and allow scientists outside of the NAWQA Program and the USGS to use these data analysis tools in their analyses.

## Purpose and Scope

The purpose of this report is to provide information on the use and capabilities of the IDAS software. This User's Manual explains how to acquire the software, load it onto a personal computer, and run the software. It discusses the relation between the NAWQA Program ecological database and the IDAS program, and provides instructions on how to use IDAS as a tool for exploring, analyzing, and exporting invertebrate data to other software programs. Though developed for processing invertebrate data downloaded from the NAWQA Program's Biological Transactional Database (Bio-TDB), the IDAS program contains tools for converting non-NAWQA data into the Bio-TDB format. Consequently, this program can be used by anyone looking for an efficient means of processing invertebrate data and generating assemblage metrics. The IDAS software was developed to help data analysts at the study-unit, regional, and national levels work either independently or in cooperation with one another, and it also facilitates the archiving of data analyses by providing procedures that automatically document the data files and options used in the analyses. The IDAS software provides a set of standard procedures for processing quantitative (richest targeted habitat, RTH, and depositional targeted habitat, DTH, samples) and qualitative (qualitative multihabitat samples, QMH) invertebrate data (Cuffney and others, 1993; Moulton and others, 2002) that are citable and that can be used by scientists within and outside of the USGS.

## Acknowledgments

The IDAS software is a distillation of 20 years of experience working with large quantities of invertebrate data in multiple databases and with numerous biologists associated with the NAWQA Program. Specifications for the software program originally were established from the input of NAWQA Program biologists during the development of the Biological Data Analysis System (BDAS) in 1994–95. The authors are indebted to the biologists who provided input at that time—Michael Bilger, Robert Black, Larry Brown, James Coles, Carol Couch, Charles Demas, Steven Frenzel, Jeffery Frey, Robert Goldstein, Steven Goodbred, Martin Gurtz, Evan Hornig, Clifford Hupp, Terry Maret, Michael Meador, Bruce Moring, Mark Munn, Karen Murray, James Petersen, Stephen Porter, Stephen Rheaume, Peter Ruhl, Barbara Scudder, Terry Short, Stephen Sorenson, Cathy Tate, Ian Waite, and Humbert Zappia. Several biologists provided input in the development of IDAS and assisted in helping to test the installation and operation of the IDAS software program—Humbert Zappia, James Coles, Ian Waite, Thomas Abrahamsen, Elise Giddings,

Robert Ourso, Mitch Harris, Terry Maret, Dorene MacCoy, Karen Murray, and Jeffrey Powell. The Ecological Integration Project and the Albemarle-Pamlico Drainage and Yakima River Basin Study Units provided additional support for the development of this software.

# Introduction to the Invertebrate Data Analysis System

The IDAS software package is a special purpose program written in Microsoft (MS) Visual Basic® 6.0. It is intended to provide NAWQA Program biologists with a flexible and efficient mechanism for analyzing invertebrate data downloaded from Bio-TDB and for preparing Bio-TDB data for use in other analytical programs, such as SYSTAT, SPLUS, MVSP, SAS, CANOCO, and Primer E. In general, the format of the invertebrate data supplied by Bio-TDB does not allow the user to analyze invertebrate data without making substantial modifications to the data files. The IDAS software program provides a quick and flexible means of manipulating and analyzing invertebrate data obtained from Bio-TDB and provides tools for importing data that are not in Bio-TDB format.

## Capabilities

The IDAS program consists of five program modules that allow the user to manipulate datasets, calculate community metrics, and export data to other programs. These modules provide the following capabilities:

1. Edit Data[1] module:

   A. Edit data files[2].

      i. Subset data, generates subsets of data based on sample information (for example, date, sample type, station) or taxonomic groupings (for example, order, family).

      ii. Combine/delete datasets:

         a. Combine tables/spreadsheets, combines MS Excel® spreadsheets or MS Access® data tables of similar structure into a new Excel spreadsheet or Access table.

         b. Delete tables/spreadsheets, deletes one or more Excel spreadsheets or Access data tables from a dataset.

      iii. Summarize taxa, identifies ambiguous taxa and provides distribution statistics for taxa in

---

[1] The bold blue type indicates the name of an IDAS module, menu, or function throughout the report.

[2] The bold blue underlined type indicates the name of a menu item throughout the report.

a dataset (number of samples and sites where the taxon occurs).

iv. Remove ambiguous parents, removes ambiguous parents from samples. Ambiguous taxa occur when abundances are reported at multiple levels of the taxonomic hierarchy in a sample (for example, Hydropsychidae, *Hydropsyche*, and *Hydropsyche sparna*). In this example, both Hydropsychidae and *Hydropsyche* are ambiguous parents of *Hydropsyche sparna*, and Hydropsychidae is an ambiguous parent of *Hydropsyche*.

v. Resolve conditional/provisional data, identifies conditional/provisional taxa (for example, *Hydropsyche* sp. nr. *simulans* Ross; Moulton and others, 2000) in the data and allows the analyst to retain the conditional/ provisional identification or assign this taxon to a related taxon (for example, *Hydropsyche* sp. or *Hydropsyche simulans* Ross).

B. Import data.

i. WR-EMAP format, imports data in a format that was used by the U.S. Environmental Protection Agency (USEPA) Corvallis Labo- ratory to store data collected as part of the Western Region Environmental Monitoring and Assessment Program (WR-EMAP).

ii. User defined format, imports data in stacked column format and converts it to Bio-TDB format (unprocessed data) based on user- supplied conversion criteria. These conver- sion criteria can be saved and used to convert additional datasets.

C. Maintain attributes file.

i. Create a new attribute file.

a. Extract taxa from an abundance file, creates a new attribute file that contains only the taxa in the abundance file and the attributes that are associated with those taxa.

> **TIP: Use the Edit Data module to select subsets of data for analysis and to examine the distribution of taxa among sites. This will help guide decisions (for example, deleting rare taxa) that are required in the Data Preparation module.**

b. Extract taxa from HIER in attribute file, creates a new attributes file that contains all the taxa that are specified in the HIER (taxonomic hierarchy) work- sheet of the attributes file. The new file contains the attributes of the taxa listed in the HIER worksheet.

ii. Modify an existing file.

a. Add taxa from an abundance file, expand an existing attributes file by adding taxa from an abundance file (in Bio-TDB format).

b. Refresh an EQTX file, refreshes an equivalent taxa (EQTX) spreadsheet based on the taxa contained in the HIER spreadsheet. This option is most often used to remove modifications to the EQTX spreadsheet that have been made previously.

c. Update the ATTRIB spreadsheet, updates the taxa attributes in the ATTRIB spreadsheet based on attributes contained in another spreadsheet. This option is most often used to incorporate information on tolerances or functional groups from other sources.

D. Utilities.

i. Convert matrix to columns, converts data in matrix format (taxa as rows, samples as col- umns or samples as rows, taxa as columns) to stacked column format. The resulting stacked count data file could then be imported using the Import data/User defined format (1,B,ii. from above) option.

ii. Random subsample, extracts a random sub- sample of data from a dataset. This option is most often used to extract a subsample based on counts (for example, 100 organism count) from a sample based on a larger count (for example, 300 organism count).

2. Data Preparation module:

A. **Select sample type(s) to process**[3], selects the sample types (ALL, QMH, DTH, RTH, and (or) QUAL = QMH+RTH+DTH) to process.

B. **Calculate density**, calculates densities as number per square meter (no./m$^2$) using the sample area information contained in the file "_Sample_All. xls" exported by Bio-TDB.

C. **Deletions based on BG processing notes**, deletes taxa based on the sample processing notes, such

---
[3] The bold magenta type indicates text associated with a frame, button, list, radio button, or check box throughout the report.

as immature or damaged, supplied by the Biological Group within the National Water Quality Laboratory (NWQL).

D. **Deletions based on lifestage**, deletes pupae and (or) terrestrial adult insects.

E. **Options based on combining lifestages**, considers or ignores lifestage (adult, pupae, or larvae) information when processing data.

F. **Options for forming qualitative (QUAL) samples**, forms QUAL samples by combining QMH, RTH, and DTH samples (QMH+RTH, QMH+DTH, or QMH+RTH+DTH).

G. **Options for processing data:**

  i. **Select lowest taxonomic level**, determines the lowest taxonomic level (for example, family, tribe, genus) at which the data will be aggregated.

  ii. **Delete rare taxa**, deletes taxa if the percentage of total abundance in a sample and (or) the percentage of sites at which the taxa occurs is less than the user specified limits.

H. **Options for resolving ambiguities:**

  i. **Resolve ambiguous taxa at or above**, removes ambiguous taxa if they occur at or above genus, tribe, subfamily, family, suborder, order, class, or phylum.

  ii. **Resolve ambiguities by**, resolves ambiguous taxa separately for each sample or for all samples combined.

  iii. **Resolve ambiguous taxa**, resolves ambiguous taxa by using one of five methods:

    a. Remove ambiguous parent and keep children (RPKC).

    b. Merge children with ambiguous parents (MCWP).

    c. Remove ambiguous parent or merge children with parent (RPMC) depending on the abundances of the children and parent.

    d. Distribute abundance of the ambiguous parent among its children (DPAC).

    e. None—retain ambiguous taxa (ORIG).

**TIP: Data must be processed by the Data Preparation module before it can be processed by the Calculate Community Metrics, Calculate Diversities and Similarities, or Data Export modules.**

3. **Calculate Community Metrics** module:

A. **Richness metrics**, calculates 25 metrics based on taxonomic richness and (or) 24 metrics based on percentage of total taxa richness.

B. **Abundance metrics**, calculates 25 metrics based on abundance of taxa, and (or) 24 metrics based on percentage of total abundance, and (or) 5 dominance metrics (percentage of total abundance represented by the most abundant taxon, the first and second most abundant, first through third most abundant, first through fourth most abundant, and first through fifth most abundant taxa).

C. **Tolerance metrics**, calculates richness and (or) abundance-weighted tolerance values along with the percentage of taxa richness and (or) abundance that was associated with a tolerance value in the attributes file.

D. **Tolerance classes**, calculates the number of taxa and percentage of total taxa richness, and (or) abundance and percentage of total abundance that fall within intolerant, moderately tolerant, and tolerant classes defined by the user on the basis of tolerance values in the attributes file.

E. **Functional group metrics**, calculates 8 richness, and (or) eight percentage richness, and (or) eight abundance, and (or) eight percent-abundance metrics based on functional group values in the attributes file.

F. **Behavioral metrics**, calculates seven behavioral metrics based on richness, and (or) seven based on percent richness, and (or) seven based on abundance, and (or) seven based on percent abundance along with the percentage of taxa richness, and (or) abundance that was assigned a behavioral characteristic in the attributes file.

4. **Calculate Diversities and Similarities** module:

A. **Diversity indices**, calculates five quantitative diversity indices, one dominance index, and three evenness indices (Appendix I).

B. **Similarity indices**, calculates two qualitative similarity indices, four quantitative similarity indices, and two quantitative dissimilarity indices (Appendix I).

5. **Data Export** module:

A. **Export data as**, exports data as abundance or density, relative (percentage) abundance or density, or proportions of total abundance or density.

B. **Transform data**, transforms data using square root, fourth root, log(X+A) or ln(X+A) before exporting.

C. **Comma delimited ASCII sample by taxa matrix**, exports data as a comma-delimited ASCII file with rows as samples and columns as taxa or rows as taxa and columns as samples.

D. **Tab delimited ASCII, full format taxa list for publications**, exports data as a tab-delimited ASCII file with rows as taxa and with the full taxonomic hierarchy and complete sample information.

E. **Stats package (with sample information)**, exports data as a comma-delimited ASCII file with sample information, but without the taxonomic hierarchy.

F. **CANOCO native format**, exports data in CANOCO condensed format for use by programs such as CANOCO or PC-ORD.

G. **Primer-E format**, exports data in a tab-delimited ASCII format that can be read by Primer-E. Includes the aggregation file based on the taxonomic hierarchy.

IDAS can process data files that are saved as Excel workbooks (2003, 2007) or Access databases (2003, 2007). All modules except for the Data Export module store data in the original files either as new spreadsheets (Excel) or new data tables (Access). The Edit Data module is the only module that can process data that are in the original Bio-TDB format, the processed format produced by the Data Preparation module, the WR-EMAP format, or user supplied format. The Data Preparation module only reads data in Bio-TDB and WR-EMAP formats. All other data formats must be converted to the Bio-TDB format in the Edit Data module before being processed in the Data Preparation module. The Data Preparation module produces a new format (processed data) that is used by the other four modules (Edit Data, Calculate Community Metrics, Calculate Diversities and Similarities, and Data Export). The Calculate Community Metrics, Calculate Diversities and Similarities, and Data Export modules can use data only in the processed format produced in the Data Preparation module. The Data Export module writes data to ASCII text files rather than back to the original Excel or Access files.

## Sources of Data Used by IDAS

All analyses are based on the invertebrate abundance (table 1) and sample area information (table 2) contained in files exported from Bio-TDB (v. 2.3.6 or newer). Invertebrate data files can be obtained from Bio-TDB by using either the **IDAS Export** (Data/Data Retrieval/IDAS Export) or the **SU Data Exports** (Data/Data Retrieval/SU Data Exports) options. Data from sources other than Bio-TDB (for example, WR-EMAP or user-defined formats) must be converted to the Bio-TDB format before processing by using the Import Data function in the Edit Data module. The **IDAS Export** option in Bio-TDB creates an Access database with tables that contain

the invertebrate data (Invert_Results_Combo), sample area information (Sample_All), and parameters used in the data retrieval (Bio_TDB_Criteria). The resulting Access database file is named by combining the identifier "IDAS" with the date and time that the data were downloaded and the user's name (for example, IDAS_20090715_1104_tcuffney.mdb[4]).

In Bio-TDB the **SU Data Exports** option allows the user to export data to either Access or Excel using the **Invertebrate Results Combined** (Data/Data Retrieval/SU Data Exports/ Invertebrate Results Combined) option to obtain the invertebrate data and the **Sample Information** option (Data/Data Retrieval/SU Data Exports/Sample Information) to obtain the sample areas. The Invertebrate Results Combined export option creates files in the form "X_Y_Z_Invert_Results_ Comb.xls," where "X" is the 4-letter abbreviation for the Study Unit, "Y" is the date (month, day, year), and "Z" is the time (hour, minute) that the data were exported (for example, ALMN_05092008_1037_Results_Comb.xls). The Sample Information must be exported if the user wants to convert abundances to densities for quantitative (RTH and DTH) samples. The **Sample Information** export option creates a separate sample all file (for example, ALMN_05122008_ 1001_Sample_All.xls; table 2) containing sample information.

The **IDAS Export** and **SU Data Exports** options of Bio-TDB create data files with slightly different arrangements of the data. The abundance data file created by the **IDAS Export** option contains the Station Name (StationName) whereas the **SU Data Exports** does not. The **SU Data Exports** option groups data together by taxa name (BU_ID) whereas the **IDAS Export** option groups data together by sample with the taxa in each sample arranged in phylogenetic order. Non-NAWQA data can be converted to Bio-TDB format by using the Import Data and Convert matrix to columns functions in the Edit Data module. If non-NAWQA data are imported with the intent of having IDAS convert these abundances to densities, then the user will have to create a "sample all" spreadsheet or table that contains the sample area in square centimeters using the formats outlined in table 2.

The Calculate Community Metrics, Calculate Diversities and Similarities, and Data Export modules use a slightly different format (processed data, table 3) as their input data. The processed data format is produced when a Bio-TDB format file is processed by the Data Preparation module. This format is similar to the Bio-TDB format but includes a column for the SUID (4-character Study Unit identifier), deletes the Notes and LabCount columns, and identifies the quantity information as abundance or density. Information on functional groups, tolerance data, and behavioral groups used to calculate community metrics is contained in the attributes file distributed with the IDAS software. This information has been compiled from USEPA (Barbour and others, 1999) and North Carolina Department of Environment and Natural

---

[4] The bold green type indicates example files that are provided with IDAS v. 5.0 and summarized in Appendix II.

**Table 1.**   Structure of the abundance data files produced by Bio-TDB using the **IDAS Export** (IDAS_20090715_1104_tcuffney.mdb) and **SU Data Exports/Invertebrate Results Combined** (ALMN_05092008_1037_Invert_Results_Comb.xls) options.

[The IDAS program uses these file formats as input to the Edit Data and Data Preparation modules. All of the columns except LabCount are used by the IDAS program. Columns in italics must be fully populated, that is, they cannot contain blank or null values. The taxonomic hierarchy associated with each BU_ID must also be populated and unique for each taxon or an error message will be generated. Note the nonstandard capitalization of SubOrder and SubFamily used by Bio-TDB]

| Column name | Data type | Example | Bio-TDB export format type | |
|---|---|---|---|---|
| | | | **IDAS Export** | **SU Data Exports** |
| *SampleID* | Long | 14102 | X | X |
| *SMCOD* | Text | ALBE0694IRM0100 | X | X |
| *STAID* | Text | 02047360 | X | X |
| StationName | Text | MILL CR | X | |
| *Reach* | Text | A | X | X |
| *CollectionDate* | Date/time | 6/1/1994 | X | X |
| Phylum | Text | Arthropoda | X | X |
| Class | Text | Insecta | X | X |
| Order | Text | Coleoptera | X | X |
| SubOrder | Text | Adephaga | X | X |
| Family | Text | Dytiscidae | X | X |
| SubFamily | Text | Hydroporinae | X | X |
| Tribe | Text | Hydroporini | X | X |
| Genus | Text | *Laccornis* | X | X |
| Species | Text | *Laccornis difformis* | X | X |
| *BU_ID* | Text | *Laccornis difformis* (LeConte) | X | X |
| *SortCode* | Long | 8002660 | X | X |
| Lifestage | Text | A | X | X |
| Notes | Text | ref. | X | X |
| LabCount | Long | 35 | X | X |
| *Abundance* | Long | 140 | X | X |

**Table 2.**   Data columns produced by the **IDAS Export** (table Sample_All in the file IDAS_20090715_1104_tcuffney.mdb) and **SU Data Exports/Sample Information** (ALMN_05122008_1001_Sample_All.xls) export options in Bio-TDB.

[Only the data columns that are used by the IDAS program to calculate densities (number per square meter) are shown]

| Column name | Data type | Example | Comments |
|---|---|---|---|
| STAID | Text | 02043500 | Station identifier |
| SampleID | Integer | 14102 | Sample identifier |
| SampleMediumCode | Text | I | Sample medium code: Invertebrate |
| SampleType | Text | R | Sample type: RTH |
| AreaSampTot | Double | 12500 | Area sampled (square centimeters) |

**Table 3.** Structure of the "processed format" spreadsheet (RTH, IDAS_20090715_1104_tcuffney.mdb) produced by the Data Preparation module.

[This file format is used as input to the other IDAS modules. It differs from the original "raw" format in that the Study Unit identifier (SUID) has been added, the Notes and LabCount columns have been deleted, and the data are expressed as density (abundance is also an option). Density is expressed as number per square meter (no./m$^2$)]

| Column name | Data type | Example | Comment |
|---|---|---|---|
| SUID | Text | ALBE | Study Unit identifier |
| STAID | Text | 02097464 | Station identifier |
| StationName | Text | MORGAN CREEK | Station name (optional) |
| Reach | Text | A | Sampling reach |
| CollectionDate | Date | 5/15/2003 | Collection date |
| SampleID | Long | 108306 | Sample identifier |
| SMCOD | Text | ALBE0503IRM0005 | Sample code |
| Phylum | Text | Arthropoda | Phylum |
| Class | Text | Insecta | Class |
| Order | Text | Ephemeroptera | Order |
| SubOrder | Text | Furcatergalia | SubOrder |
| Family | Text | Ephemerellidae | Family |
| SubFamily | Text | | SubFamily |
| Tribe | Text | | Tribe |
| Genus | Text | *Dannella* sp. | Genus |
| Species | Text | *Dannella simplex* | Species |
| BU_ID | Text | *Dannella simplex* (McDunnough) | Taxon name |
| SortCode | Long | 87 | Sort code |
| Lifestage | Text | L | Lifestage |
| Density | Double | 2.4 | Density (no./m$^2$) |

Resources (2006) data. Versions of the attributes file are identified by appending the version number to the attribute file name (for example, the current version is named Attributes_BEHAV_v5a.xls).

The IDAS software is designed to work with Microsoft Access (*.mdb, *.accdb) and Excel files (*.xls and *.xlsx) provided that the formats of the Access files are the same as the original Excel files obtained from Bio-TDB, the WR-EMAP format, or are converted to the Bio-TDB format using the Utilities and Import Data tools provided in the Edit Data module. The user does not need to create "keys" when creating Access files because IDAS will automatically create the keys that it requires. While IDAS will read and write data in Excel files, the use of Access is highly recommended because it enforces strict data typing and can substantially increase processing speed compared to Excel files. The attributes files (Attributes_BEHAV_v5a.xls) should not be converted to an Access file because the design of the IDAS software expects this information to be contained in an Excel file.

## Characteristics of Bio-TDB Data

Invertebrate abundance data exported from Bio-TDB are structured to provide a maximum amount of information in a minimum amount of space. These files consist of 20 (**SU Data Exports**) or 21 (**IDAS Export**) columns of data that provide information on the identity of the sample (Sample Identifiers), the taxonomic hierarchy associated with each taxon (Taxonomic Hierarchy), and the abundance of organisms in the sample (table 1). Data are presented in stacked column format sorted in descending phylogenetic order (SortCode) by sample (**IDAS Export**) or for the entire dataset (**SU Data Exports**). The stacked column format places information for a taxon (abundance, sampling information, taxonomic hierarchy) into columns and places taxa for a sample in consecutive rows of data stacked one below the other. This minimizes the size of datasets compared to a matrix format (for example, taxa as rows, samples as columns), but it cannot represent a sample that does not have any taxa because there is no provision to represent an abundance of zero.

The Bio-TDB format preserves information on lifestage (adult, pupa, larva) and sample processing notes (Notes) that are associated with sample components and processing fractions (Moulton and others, 2000). Therefore, a single taxon (BU_ID) in a sample may have multiple rows of information (table 4) representing unique combinations of BU_ID, lifestage, and processing notes. This ensures that all the information generated by the National Water Quality Laboratory Biological Group (NWQL BG) during sample processing is available to the analysts as they decide how to summarize their data. Consequently, data exported from Bio-TDB require some preparation and manipulation before the data can be used to calculate community metrics or as input to other analysis programs.

The Bio-TDB export options also produce an Excel workbook or Access data table that contains the area sampled in square centimeters for the quantitative invertebrate samples (RTH and DTH). The suffix "_Sample_All" is used by Bio-TDB to identify the workbook or data table that contains the sample areas. The data in the Bio-TDB abundance file represents the numbers of organisms in the samples and can be converted to densities (number per square meter) using the sample areas contained in the "_Sample_All" file or table. The "_Sample_All" file or table contains 41 columns of information; however, the IDAS program uses only 5 columns of data (table 2) to match information on the area sampled with abundance data for each quantitative (RTH, DTH) sample. The calculation of densities is optional in the IDAS program.

## Provisional and Conditional Identifications

When a specimen cannot be identified by the NWQL BG down to the taxonomic level specified in the sample processing protocol (Moulton and others, 2000), the presence or abundance of the specimen usually is reported at a taxonomic level that is higher than the target level (for example, genus instead of species or family instead of genus). However, the NWQL BG occasionally will assign a provisional or conditional identification to a specimen. This occurs when (1) the specimen represents a potentially undescribed species (*Hydropsyche* sp. nr. *simulans*), (2) a species differs in some minor way from the description in the literature (*Hydropsyche* cf. *simulans*), (3) one of two taxa cannot be resolved (*Hydropsyche rossi/simulans*), (4) a taxon is provisional in the literature (*Oecetis* sp. A), (5) a group of closely related species cannot be separated (*Hydropsyche scalaris* group), or (6) some other cause results in the use of a nondefinitive identification. These provisional and conditional identifications are only present in the BU_ID column of the data exported from Bio-TDB; they are not propagated in the taxonomic hierarchy, which contains only definitive identifications. For example, the BU_ID column in table 5 contains the conditional species

**Table 4.** Example of multiple entries for one taxon associated with a single sample.

[The multiple entries represent unique combinations of BU_ID, lifestage, and notes. In this way, Bio-TDB preserves all information generated during the processing of each invertebrate sample component]

| SampleID | BU_ID | Lifestage | Notes | Abundance |
|----------|-------|-----------|-------|-----------|
| 13866 | *Hydroptila* sp. | L | ref. | 1 |
| 13866 | *Hydroptila* sp. | P | | 5 |
| 13866 | *Hydroptila* sp. | L | imm. | 50 |
| 13866 | *Hydroptila* sp. | L | dam. | 62 |
| 13866 | *Hydroptila* sp. | L | dam., imm. | 65 |

*Hydropsyche betteni/depravata*, but the lowest taxonomic level reported in the taxonomic hierarchy for this taxon is the genus (*Hydropsyche*). Non-NAWQA data that will be imported into IDAS should identify provisional/conditional taxa as described in table 5 and should not propagate provisional/conditional identifications in the taxonomic hierarchy in order to ensure that data are processed as described in this manual.

Provisional and conditional identifications can provide the analyst with additional information that may be of use in understanding the distribution of invertebrates. However, including these identifications may not be appropriate for some types of analyses. Because the taxonomic hierarchy in the Bio-TDB data contains only unequivocal identifications, provisional and conditional identifications can be eliminated simply by setting the lowest taxonomic level to something other than BU_ID (for example, Species). This may produce datasets that differ from the original dataset even when the lowest level of the taxonomic hierarchy (species) is used. Table 5 illustrates how provisional and conditional identifications disappear when the original BU_IDs are converted to one of the levels in the taxonomic hierarchy. For example, at the species level *Hydropsyche betteni/depravata*, *Hydropsyche* cf. *simulans*, and *Hydropsyche* sp. A all become *Hydropsyche*; *Bezzia/Palpomyia* becomes Ceratopogonidae; *Cricotopus bicinctus* group becomes *Cricotopus*; and *Stilocladius*? becomes Chironomidae.

## Ambiguous Taxa

Invertebrate data downloaded from Bio-TDB may contain taxonomic ambiguities within and among samples. Taxonomic ambiguities occur when specimens cannot be identified to the level of taxonomic resolution specified in the sample-processing protocol (Moulton and others, 2000) and must be reported at higher taxonomic levels (for example, family rather than genus). If these specimens are parents of other taxa in the sample or dataset, then the sample or dataset contains ambiguous taxa. Table 6 presents a hypothetical sample containing ambiguities in the mayfly family Baetidae. This sample contains 144 specimens identified to species, 20 identified to genus, and 200 identified to family. The three

**Table 5.**  Conditional and provisional identifications are confined to the BU_ID column of the Bio-TDB export file.

[Conditional and provisional identifications are indicated by the appearance of "nr ," "cf.," "/," "group," "complex," "n. sp.," or "?" in the taxonomic designation or by a species or genus name that consists of a single letter or number (sp. 1, sp. 2, or Genus A; see Moulton and others, 2000). Provisional and conditional identifications are not propagated in the taxonomic hierarchy]

| Original BU_ID | Taxonomic hierarchy | | | |
|---|---|---|---|---|
| | Species | Genus | Family | Order |
| Glossomatidae | | | Glossomatidae | Trichoptera |
| *Agapetus* | | *Agapetus* | Glossomatidae | Trichoptera |
| *Glossosoma* | | *Glossosoma* | Glossomatidae | Trichoptera |
| Hydropsychidae | | | Hydropsychidae | Trichoptera |
| *Hydropsyche* | | *Hydropsyche* | Hydropsychidae | Trichoptera |
| *H. betteni* | *H. betteni* | *Hydropsyche* | Hydropsychidae | Trichoptera |
| *H. betteni/depravata* | | *Hydropsyche* | Hydropsychidae | Trichoptera |
| *H.* cf. *simulans* | | *Hydropsyche* | Hydropsychidae | Trichoptera |
| *H.* sp. A | | *Hydropsyche* | Hydropsychidae | Trichoptera |
| *Ceratopsyche* | | *Ceratopsyche* | Hydropsychidae | Trichoptera |
| *Bezzia/Palpomyia* | | | Ceratopogonidae | Diptera |
| *Cricotopus bicinctus* group | | *Cricotopus* | Chironomidae | Diptera |
| *Stilocladius?* | | | Chironomidae | Diptera |

species are ambiguous children of ambiguous parents *Baetis* and Baetidae. Baetidae is the ambiguous parent of *Baetis*.

Ambiguous taxa present a problem in the analysis of invertebrate data, particularly in the determination of taxa richness and abundance metrics. For example, what is the taxa richness represented in table 6? Some biologists would argue that there are only three taxa (species) in this dataset; others would argue that there are five taxa. Assuming there are only three taxa (species), what becomes of the abundance associated with Baetidae and *Baetis*? If the user assumes that there are five taxa, then the information on the richness of Baetidae and *Baetis* is superfluous given that they already are represented as parents of the three species. Consequently, the manner in which ambiguous taxa are resolved can have a significant effect on the comparability of data and interpretation of results (Cuffney and others, 2007).

The purpose of the IDAS program is not to advocate a particular approach to resolving the problem of ambiguous taxa, but rather to provide analysts with a set of tools that will

allow them to process these data in an efficient and consistent manner according to the decisions that they deem appropriate to their analysis. The IDAS program has eight options (in the Data Preparation module) for resolving ambiguous taxa either on a sample-by-sample basis or for a group of samples. The IDAS program can resolve ambiguous taxa in large datasets and has been tested on datasets with over 400 samples and more than 700 taxa.

## Installation

The IDAS program is a stand-alone program designed to run on a laptop or desktop computer with a MSWindows® operating system and MS Excel installed locally. The IDAS software can be obtained on the Web (*http://nc.water.usgs. gov/usgs/albe/idas/*), by anonymous ftp (*ftp://ftpext.usgs.gov/ pub/er/nc/raleigh/tfc/IDAS_v5/*) or by contacting the author for a compact disc (CD). The installation CD contains the IDAS program, user's manual, support files (invertebrate attributes files), examples of Bio-TDB export files, and examples of non-NAWQA data files that can be imported into IDAS. Even though the IDAS software was written specifically to work with NAWQA Program data downloaded from Bio-TDB, it contains several tools that facilitate the conversion of non-NAWQA datasets into Bio-TDB format. While it is possible to manually convert data into Bio-TDB format, it is recommended that data be imported using the data import functions in the Edit Data module because these functions check for common data conversion and formatting errors.

**Table 6.**  Example of ambiguous taxa.

[Both *Baetis* and Baetidae are ambiguous parents of the three species. Baetidae is an ambiguous parent of *Baetis*. The three species are ambiguous children of the genus *Baetis* and family Baetidae]

| Family | Genus | Species | Abundance |
|---|---|---|---|
| Baetidae | | | 200 |
| | *Baetis* | | 20 |
| | | *Baetis bicaudatus* Dodds | 34 |
| | | *Baetis brunneicolor* McDunnough | 65 |
| | | *Baetis intercalaris* McDunnough | 45 |

## System Requirements

The following computer hardware and software are required for the operation of the IDAS program.

**Processor:** Pentium® or compatible microprocessor running at 90 megahertz (MHz) or higher.

**Hard disk space:** approximately 35 megabytes (MB) for installation, although the installed program requires only about 14 MB of disk space.

**System memory:** a minimum of 64 MB of random access memory (RAM), although 128 MB or more is preferred.

**Video:** a minimum of 800 by 600 pixels, 1,024 by 768 preferred. The program screens are sized to run on a laptop.

**Mouse, touch pad, or other pointing device:** required.

**Software:** MS Excel version 8.0 or later required. The IDAS program cannot save files in MS Excel format unless Excel is loaded on the user's computer.

**Operating system:** MS Windows XP® or Windows Vista® is required. Users desiring to install IDAS on computers running earlier operating systems should consult Microsoft's Web site (*http://www.microsoft.com*) for information on updating these operating systems.

**Administrator rights:** Installation of IDAS on USGS computer systems requires administrator rights.

## Installing the IDAS Software

The installation package for the IDAS software can be obtained via anonymous ftp from the North Carolina Water Science Center (*ftp://ftpext.usgs.gov/pub/er/nc/raleigh/tfc/IDAS_v5/Installation/*) or by contacting the author for a CD copy of IDAS. The installation package consists of four files that can be obtained separately or combined into a single compressed file (Idas.zip):

1. IDAS.CAB – a compressed file that contains the IDAS program, forms, and support files as packaged by the Visual Basic software packaging and distribution tool.

2. setup.exe – a small program that installs the IDAS software (IDAS.CAB) on the host computer.

3. SETUP.LST – a file that contains setup information used by setup.exe.

4. IDAS_Defaults.txt – a text file that stores the default settings for the IDAS program.

5. IDAS.ZIP – a compressed file (zip file) that contains all four of the above files.

6. IDAS.exe – the IDAS program without the supporting default and installation files.

Installation begins by copying the zip file (IDAS.ZIP) or the individual installation files (IDAS.CAB, setup.exe, SETUP.LST, and IDAS_Defaults.txt) into a temporary directory on the host computer. The IDAS program uses a standard MS Windows installation program that copies the program files to new directories, registers the various program files, and adds IDAS to the startup menu. By default, the installation program installs IDAS and supporting files in the **Program Files** directory of the boot drive in a group called **EcoTools**, although the user can specify a different directory during the installation process.

Installation of IDAS on USGS computer systems requires that the user be logged on with administrative rights. Installation begins by running the set-up program (setup.exe) after closing all other programs. Setup.exe is run by using the **Start** plus **Run** options on the main Windows screen and following the instructions provided by the installation program. The IDAS software will be installed in a folder called **Invertebrate Data Analysis System** within the **Programs** folder. The installation program also will create an **EcoTools** directory that can be used to hold data and output files. In the event that the installation program signals an alert that it is attempting to replace a newer version of a support file with an older version, select the option to keep the newer version of the support file. Once the installation process is completed, the files in the temporary directory can be deleted.

> **CAUTION:** The installation package will alert the user when it tries to replace an existing file with a version that is older. The user will be given the option of replacing the newer file with the older version or keeping the newer version. ALWAYS keep the newer version!

## Updates

The IDAS software and documentation are updated periodically, on the basis of requests from users for new features or the discovery of bugs in the software. The latest versions of the software and documentation are available on the anonymous ftp site. Users do not have to go through the entire installation process each time the IDAS software is updated. The IDAS software can be updated simply by downloading the latest version of the program executable file (IDAS.exe) from the IDAS ftp site and copying it over the previous copy of IDAS.exe in the folder "Programs/Invertebrate Data Analysis System" or wherever the program was installed.

## Help and Documentation

The primary sources for help with installing and using the IDAS software are the user's manual, the program developer, and other IDAS users. The IDAS program does not have a fully implemented help system. Help within the IDAS program is limited to explanatory text that appears when the cursor is held over certain selection buttons or boxes. The following sources provide information and documentation that may be useful to IDAS users.

Email the program developer at:
*tcuffney@usgs.gov*

Electronic copies of the IDAS program and documentation can be obtained from:
*ftp://ftpext.usgs.gov/pub/er/nc/raleigh/tfc/IDAS_v5/*

Information on Bio-TDB, including documentation, can be obtained from:
*http://nc.water.usgs.gov/usgs/biotdb/*

Information on MS Excel and Access can be obtained from the appropriate user manuals or online help:
*http://www.microsoft.com*

Information on invertebrate tolerances can be obtained from:
Barbour, and others (1999) or *http://www.epa.gov/ owow/monitoring/rbp/*
North Carolina Department of Environment and Natural Resources (2006) or *http://www.esb.enr. state.nc.us/BAUwww/benthossop.pdf*

Information on diversity and similarity indices can be obtained from:
Brower and Zar (1984) and Washington (1984)

Information on the National Water-Quality Assessment (NAWQA) Program can be obtained from:
*http://water.usgs.gov/nawqa/*

Suggested reference for IDAS:
Cuffney, T.F., and Brightbill, R.A., 2010, User's manual for the National Water-Quality Assessment Program Invertebrate Data Analysis System (IDAS) software, version 5: U.S. Geological Survey Techniques and Methods 7–C4, 126 p.

# Using IDAS

The IDAS program is designed to be a single-user system; that is, it will not open files for simultaneous use by multiple users or programs. This was done to protect the integrity of the data files and to simplify construction of the program. Consequently, the user should avoid using other programs, such as MS Excel® or Access,® to view data or attribute files while IDAS is running. If another program attempts to access a file that is already in use by IDAS, an error will be generated and data processing will cease. In addition, IDAS uses a hidden copy of Excel to save data to Excel workbooks. Starting Excel while the hidden copy is running will create a second instance of Excel, which can substantially reduce the memory available to IDAS and substantially reduce program performance. To avoid possible conflicts, it is best not to run Excel or Access while using IDAS or to limit the use of these programs to the opening window of IDAS or the opening windows of the individual modules. If Excel or Access is used in this manner, be sure to exit these programs before returning to the IDAS program.

## Starting IDAS

There are four methods to start the IDAS program once it has been properly installed. While all of these methods work, the first method is the most convenient for starting IDAS, particularly if the IDAS program is used frequently.

1. Create a program shortcut by using Windows Explorer® to view the IDAS program (IDAS.EXE) and then dragging the IDAS icon onto the Windows Desktop.® Right click on the resulting icon and rename it "IDAS," if desired. IDAS can now be started by double clicking the Desktop icon.

2. Use the **Start** button in MS Windows.® The sequence is Start/Programs/Ecotools/IDAS.

3. Double click the program name from within Windows Explorer.

4. Use **Start** and **Run** options in MS Windows. The full name of the executable file must be entered (that is, include the path) in the "Run" window or the file can be selected by using the "Browse" button.

Once IDAS has been started, the opening screen will appear (fig. 1). This screen has six buttons, one for each of the five data-processing modules and an Exit button that allows the user to terminate IDAS. The Options menu allows the user to set program options such as the default input data type (Excel or Access), how to list the children of ambiguous parents (by occurrence or alphabetically), how to determine the identifier (SampleID or STAID) used for samples in

> **TIP: DO NOT start Excel or Access while IDAS is running except at the opening window of IDAS or one of its modules. Be sure to close Excel or Access before returning to IDAS.**

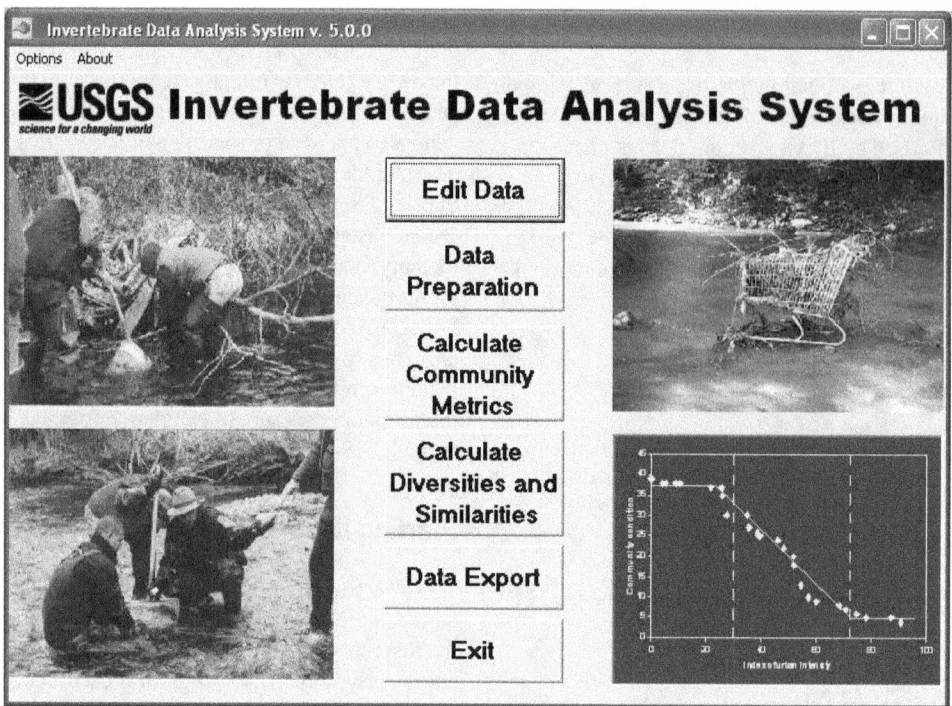

**Figure 1.**   Opening screen of the IDAS program showing the buttons that activate the five program modules.

Primer output, and settings for tolerance class ranges (fig. 2). Default options are selected by clicking on the appropriate radio buttons and then clicking on the **Apply** button. The values that are currently in effect are saved as the default values in the file IDAS_Defaults.txt and are automatically loaded whenever IDAS is started. The Options menu is also displayed in the opening window of each of the five data processing modules. The About menu summarizes the features of the IDAS program and provides contact and support information (fig. 3). Each module has its own About menu item that helps the user understand what each module can do. Clicking on a button starts the corresponding data-processing module.

## Common Features of Modules

The five IDAS modules have many common features, such as mechanisms for opening and closing files, selecting tables or spreadsheets, viewing data, displaying error messages, and providing information on program modules and contact information. Data processing in each module starts with the user selecting an Excel or Access file that contains the data needed by the program. All modules except the Export Data module save data in the Excel workbook or Access database that provided the invertebrate abundance data.

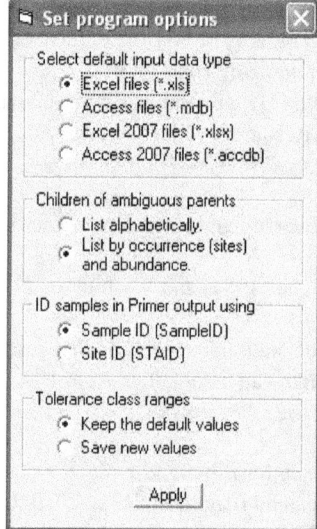

**Figure 2.**   The Set program options window allows the user to set the program defaults.

[Program defaults include input file types, how children of ambiguous parents are listed, how samples are identified in Primer output, and tolerance class ranges that determine intolerant, moderately tolerant, and tolerant classes]

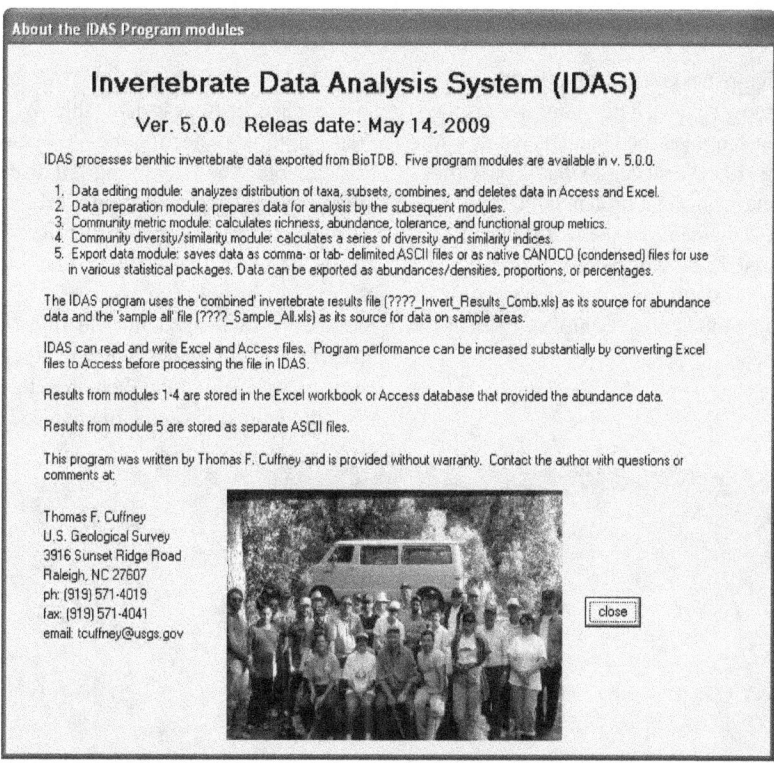

**Figure 3.** The About window tells the user what the program or module is capable of doing.

[Program defaults include input file types, how children of ambiguous parents are listed, how samples are identified in Primer output, and tolerance class ranges that determine intolerant, moderately tolerant, and tolerant classes]

## Menu Items

The Files menu is used to open and close data files in all modules except the Edit Data module. The Open option is used to open an Excel or Access file that contains the abundance data. The Close option is used to close all open files, reset the module, and prepare it for processing another set of data (in contrast to Exit). In the Edit Data module, the Files option leads to a series of submenus that are used to select specific data-editing features and the files that are appropriate for the editing features and gives the user the option to import data using the WR-EMAP data format or a user-defined format. The View menu is used to display the first 20 lines of the spreadsheet or data table that is providing abundance data to the program. In the Calculate Community Metrics module, the View menu also can be used to view and print a list of metrics calculated by IDAS. The Run menu processes data according to the options selected by the user. The Exit menu is used to exit the current module and return the user to the opening window of IDAS. The Close menu prepares the module for processing another dataset by resetting the module without exiting the module. The user also can exit the module by clicking on the "x" in the upper right-hand corner of the module window. The Options menu (fig. 2) is displayed in every module and can be modified as needed. The About menu (fig. 3) displays a window that provides a synopsis of the module capabilities and contact information. Menu items are activated at appropriate points in the program; for example, the Close option in the Files menu is inactive (that is, dimmed) until a file has been opened.

> **TIP: Data must be processed by the Data Preparation module before metrics or indices can be calculated.**

## Status Bars

Each module has a five-panel **status bar** along the bottom of the module window (fig. 4). This status bar displays (1) the name of the file that provided the abundance data and its path, (2) the type of file (Excel or Access) that is providing the data, (3) the name of the worksheet or data table that contains the source data, (4) the name of the worksheet or data table where the processed data will be stored, and (5) user prompts and program status messages (for example, "Finished" indicates that this module has completed processing data).

## Loading Data

Data are loaded into the IDAS program from Excel spreadsheets or Access data tables by using a two-step process. Step 1 involves selecting the file that contains the spreadsheet or data table. The IDAS program displays a standard Window interface (fig. 5) for selecting files. Excel files are the default setting of the file-selection interface; however, Access files can be viewed and selected by clicking on the down arrow of the Files of type text box which is part of the main IDAS file selection window, and selecting "Access files (*.mdb or *.accdb)." A file can be selected either by clicking on a file name to highlight it and then clicking on the Open button or by double clicking on the file name.

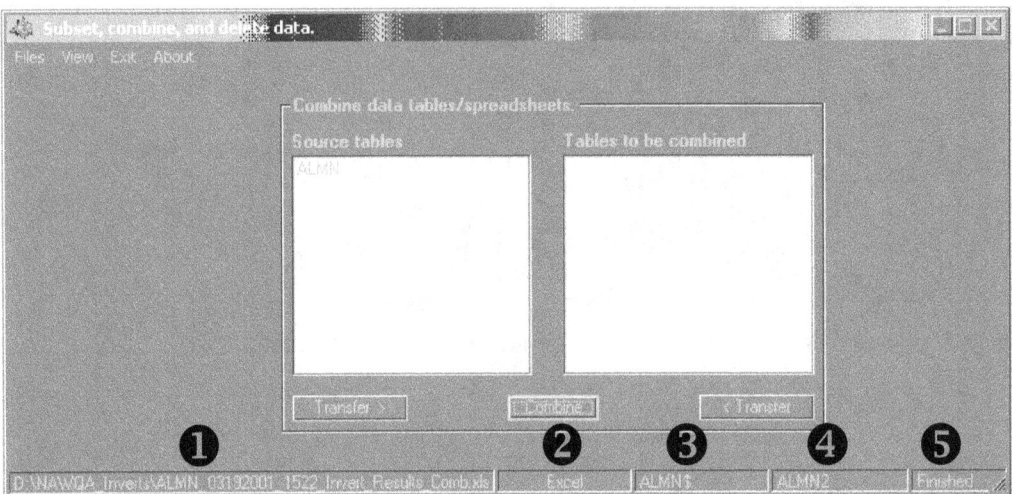

[The status bar shows the (❶) source file name, (❷) file type, (❸) source data table/spreadsheet name, (❹) destination data table/spreadsheet name, and (❺) user prompts and status messages]

**Figure 4.** A five-panel status bar displays information at the bottom of each module window.

TIP: The source file name panel expands as the size of the file name increases. Keep the source file name and path short so that other information will be visible in the status bar.

[Excel files are displayed by default. Access files may be viewed and selected by clicking on the down arrow of the Files of type text box and selecting "Access files (*.mdb)." The default file type can be changed by setting the program options (fig. 2)]

**Figure 5.** File-selection window displayed by IDAS modules for opening data files.

Step 2 involves selecting a spreadsheet or data table that contains data. All IDAS modules display the same window (fig. 6) for selecting a spreadsheet or data table. IDAS displays all the spreadsheets or tables that correspond to the data format required by the module. The user selects a spreadsheet or table either by clicking on a name to highlight it and then clicking on the Select button, or by double clicking the name. The Cancel button will reset the module and return the user to the opening screen of the module.

Every module checks the structure of the data files (Bio-TDB and processed data) every time they are opened. The program checks that the required columns of data are present and that the data columns are correctly populated. The SUID, STAID, SMCOD, SampleID, Reach, CollectionDate, BU_ID, and Abundance columns must be fully populated (no missing values). If the program detects missing values, it stops processing the data and will reset the module after displaying an error message that lists the data columns that are missing values (fig. 7). The taxonomic hierarchy also must be correctly populated, that is, the columns phylum to species cannot all be blank, and the hierarchy must be unique for each BU_ID at each level of the taxonomic hierarchy. If the program detects a problem with the taxonomic hierarchy, it will generate an error message (fig. 8) and give the user the option of exiting the program and manually correcting the taxonomic hierarchy or having the program automatically correct problems with the hierarchy. If the user elects to have the program automatically correct problems with the taxonomic hierarchy, IDAS will chose the taxonomic hierarchy that is most frequently associated with the BU_ID. This may or may not be the desired taxonomic hierarchy. Each BU_ID also must have a SortCode that is unique for each BU_ID. If the program detects missing sort codes, multiple sort codes associated with a BU_ID, or multiple BU_IDs associated with a SortCode, it will provide a message alerting the user of the problem (fig. 9) and will provide the option of creating new SortCodes or continuing

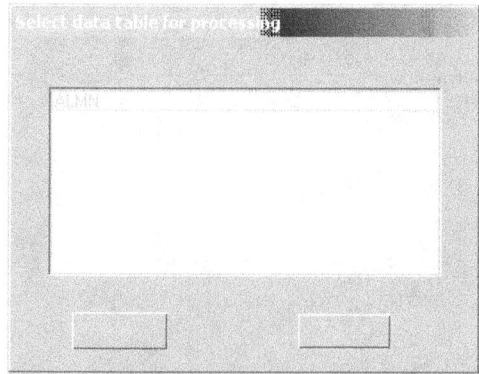

**Figure 6.**   Standard window displayed by IDAS for selecting an Excel spreadsheet or Access data table.

[A data table or spreadsheet can be selected by clicking on the name to highlight it and then clicking on the "Select" button or by double clicking on the name]

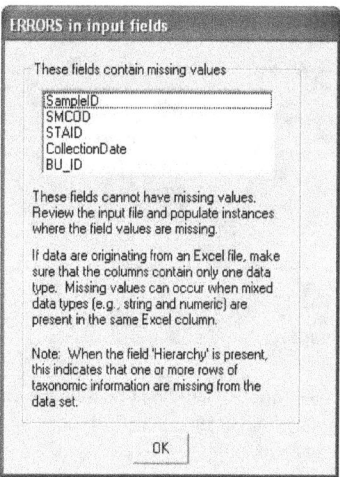

**Figure 7.**   The IDAS program checks for missing values in the invertebrate data file.

[SUID, STAID, SMCOD, SampleID, Reach, CollectionDate, BU_ID, and Abundance columns cannot have missing values. The taxonomic hierarchy must also be populated. Missing data must be corrected outside of the IDAS program]

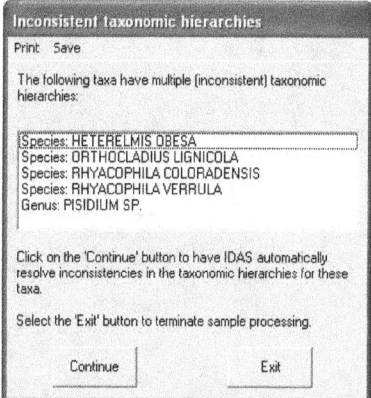

**Figure 8.**   The IDAS program checks for consistency in the taxonomic hierarchy associated with each BU_ID and reports the name and taxonomic level at which the inconsistency occurs.

[The names and taxonomic levels at which the inconsistancies occur are reported to the user. This information can be printed (Print) or saved to an ASCII text file (Save). Taxonomic inconsistencies can be corrected outside of the IDAS program (Exit) or automatically within the IDAS program (Continue). Automatic correction chooses the taxonomic hierarchy that is most frequently associated with the BU_ID]

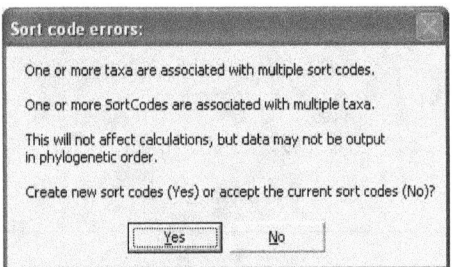

**Figure 9.** The IDAS program checks SortCodes to ensure that they uniquely identify each BU_ID.

[If the SortCodes are incomplete or not unique, the user is given the option of generating new SortCodes or proceeding with the current sort codes]

with the current SortCodes. SortCodes are only used to sort data into phylogenetic order and are not critical to the processing of the invertebrate data. SortCodes are created by sorting the taxonomic hierarchy according to the sequence outlined in table 7.

Many of the IDAS modules automatically save output to spreadsheets or tables by appending a standard suffix (for example, "_Diversity") to the name of the spreadsheet or data table (for example, diversity values derived from NoAmbig are stored in NoAmbig_Diversity). This helps document the analysis by linking the output data table or spreadsheet with

the source data table or spreadsheet. After the user has selected a spreadsheet or data table as an input source, the IDAS program will automatically scan to see if executing the module will produce a duplicate spreadsheet or data table name in the Excel workbook or Access database. If IDAS discovers a duplicate output name, it will display a window that warns the user that duplicate data tables or spreadsheets exist (fig. 10) and lists the duplicate file names. The user has the option of overwriting existing tables (Overwrite button) or terminating (Cancel button) the procedure. Clicking on the Cancel button will reset the module, which gives the user the opportunity to change the name of the existing spreadsheet or data table so existing information is not lost by overwriting the spreadsheet or data table with new data.

## Resetting or Exiting a Module

The user can reset a module by selecting the Close option from the Files menu. This will return the user to the opening window of the module and prepare the module to accept a new dataset. This is the procedure to follow if the user wishes to process multiple datasets through the same module. The user can exit a module by selecting Exit from the menu bar. This will close the module and return the user to the opening screen of the IDAS program (fig. 1). Selecting the Exit button from the opening window of IDAS will close the IDAS program. The user may also exit a module or the IDAS program by clicking on the "x" located on the upper right-hand corner of the active window.

**Table 7.** SortCodes are formed by sorting the taxonomic hierarchy associated with each BU_ID from phylum to suborder according to the entries in the sequence (SEQ) column.

[Families, subfamilies, tribes, genera, and species are sorted in alphabetical order]

| SEQ | Phylum | Class | Order | SubOrder |
|---|---|---|---|---|
| 1 | Porifera | Turbellaria | Decapoda | Nematocera |
| 2 | Cnidaria | Gastropoda | Mysida | Brachycera |
| 3 | Platyhelminthes | Bivalvia | Isopoda | |
| 4 | Nemertea | Polychaeta | Amphipoda | |
| 5 | Gastrotricha | Aphanoneura | Collembola | |
| 6 | Rotifera | Oligochaeta | Ephemeroptera | |
| 7 | Nematoda | Brachiobdellida | Odonata | |
| 8 | Nematomorpha | Hirudinea | Plecoptera | |
| 9 | Tardigrada | Arachnida | Orthoptera | |
| 10 | Bryozoa | Branchiopoda | Hemiptera | |
| 11 | Mollusca | Maxillopoda | Megaloptera | |
| 12 | Annelida | Ostracoda | Neuroptera | |
| 13 | Arthropoda | Malacostraca | Trichoptera | |
| 14 | | Insecta | Lepidoptera | |
| 15 | | | Coleoptera | |
| 16 | | | Diptera | |

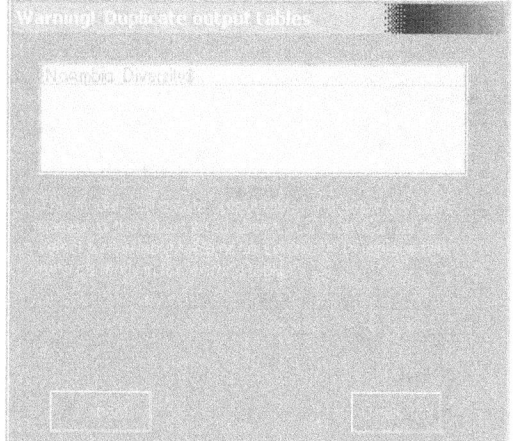

TIP: Worksheet names can be changed in Excel by double-clicking on the worksheet tab. Access tables can be renamed by right-clicking on the table name in Access.

**Figure 10.**   The IDAS program warns the user if executing the module will produce a spreadsheet or data table name that already exists in the Excel workbook or Access database.

## Automatic Documentation

The IDAS program provides automatic documentation for files that are created or modified using the Edit Data module and abundance data processed by the Data Preparation, Calculate Community Metrics, and Calculate Diversities and Similarities modules. Changes to the invertebrate attributes file (Attributes_BEHAV_v5a.xls) are documented in the "Version" spreadsheet, which stores information on the data sources and actions that were used to modify or create the attributes workbook (table 8). The "Modified" columns in the HIER and EQTX spreadsheets of the attribute file document records when information for specific taxa were created or modified. The "Notes" spreadsheet documents the data sources and processing steps used to create the original attribute workbook (Attributes_BEHAV_v5a.xls) included with IDAS. This spreadsheet can be manually updated to record changes made by the user.

Documentation for the processing of the invertebrate abundance data begins when the Data Preparation module creates an "options" spreadsheet or table by appending the suffix "_Options" to the name of the processed data (for example, RTH_Options). The "_Options" spreadsheet or table is used to store information on the data sources (that is, files and tables or spreadsheets supplying abundances, sample areas, and attributes), the processing options that were selected, output table or spreadsheet names, file names (if new files were created), and the date and time that the processing occurred (table 9). If the user creates a processed data spreadsheet or table manually (for example, extracts a subset of data using Access rather than using the options in the Edit Data module), the "options" spreadsheet or table will not be created. If IDAS detects that the "options" spreadsheet or table is missing, it will alert the user that the documentation procedures are no longer operational and provide the option of proceeding without documentation.

**Table 8.**   The "Version" spreadsheet in the attributes file documents the data sources and processing options used to create the attributes file.

[In this example the Attributes_BEHAV.xls file created on 12/10/2006, which was originally optimized for NAT_TOL, was optimized for SE_TOL on 8/4/2008 based on the taxa in the abundance spreadsheet "Data" of the file D:\ALBE_Invert.xls. The newly created attributes spreadsheet stores the new Attribute and Version information]

| Created | Attribute_source_file | Tolerance_source | Abund_source_file | Abund_table |
|---------|----------------------|------------------|-------------------|-------------|
| 12/10/2006 10:33 | D:\Attributes_BEHAV.xls | Nat_TOL | | |
| 8/4/2008 15:06 | D:\Attributes_BEHAV.xls | SE_TOL | D:\ALBE_Invert.xls | Data |

**Table 9.**    The "_Options" spreadsheet or table documents the program settings, data sources, and output files used or generated by the IDAS program

[The "_Options" spreadsheet or table is created by the Data Preparation module and updated by other modules. If the IDAS program detects that the "_Options" spreadsheet or table is missing, it will alert the user that it is missing and that the operations are not being documented]

| Options | Selected |
|---|---|
| Source file for abundance data | D:\ALBE_Inverts.xls |
| Calculate densities | Yes |
| Source file for area sampled | D:\ALBE\Inverts_Sample_All.xls |
| Sample type(s) selected for processing | RTH |
| Delete artifacts | NONE |
| Delete immatures | NONE |
| Delete damaged specimens | NONE |
| Delete specimens with wrong gender for identification | NONE |
| Delete specimens with indeterminant identifications | NONE |
| Delete specimens where poor mounts interfere with IDs | NONE |
| Delete pupae | No |
| Delete terrestrial adults | Yes |
| Keep lifestages separate | No |
| Combine lifestages for each BU_ID | Yes |
| Lowest taxonomic level allowed | BU_ID |
| Combine samples before resolving ambiguities | Yes |
| Option 1: Drop parents, keep children | No |
| Option 2: Add children to parents | No |
| Option 3: Keep children if greater than parents | No |
| Option 4: Distribute parents among children | Yes |
| Option 5: None – do not resolve ambiguities | No |
| Tolerance data options | Date created: 6/7/2005 |
| Source file for abundance data | D:\ALBE_Inver.xls |
| Source spreadsheet for abundance data | RTH_DPAC |
| Source file for tolerance data | C:\Attributes_BEHAV_SE_TOL.xls |
| Regional source of tolerance data | SE_TOL + Nat_TOL |
| Tolerance metrics saved to | RTH_DPAC_TOL_Metrics |

## Recommended Workflow

The IDAS program provides a variety of options for processing invertebrate data, calculating metrics, diversity and similarity indices, and exporting to other software packages. The following workflow is recommended as the basis for processing data through the IDAS program. This workflow can be adjusted as the user gains experience with the program and how it manipulates data.

1. **Assemble the data to be analyzed:** Use the functions in the Edit Data module to import non-NAWQA data, extract subsets of data for analysis (RTH or QMH), or to combine datasets from multiple spreadsheets (Excel) or data tables (Access). If datasets reside in separate Excel workbooks or Access databases, the datasets will have to be combined into one workbook or database before they can be manipulated in IDAS. The objective is to get all the data that will be analyzed together into a single workbook or database.

2. **Summarize the distributions of taxa:** Use the Summarize taxa function in the Edit Data module to examine the distribution of taxa among sites and samples and to identify ambiguous parents and children. This analysis is particularly useful for reviewing the taxonomy of the dataset and for making decisions about how to handle rare taxa and resolve ambiguous and provisional/conditional taxa.

3. **Resolve provisional/conditional taxa:** If desired, use the Resolve conditional/ provisional taxa function in the Edit Data module to change the taxonomic identifiers of conditional/provisional taxa.

4. **Process data through the Data Preparation module:** Use the Data Preparation module to convert abundances to densities, select sample types to process, make deletions based on laboratory processing notes and lifestages, set lowest taxonomic levels, delete rare taxa, and resolve ambiguous parents.

5. **Summarize the distributions of taxa:** Use the Summarize taxa function in the Edit Data module to examine how processing data through the Data Preparation module has affected your data. For example, did the processing options that were selected remove rare taxa and ambiguous taxa?

6. **Create an attribute file that is optimized for the data:** Use the Maintain attributes file functions in the Edit Data module to create an attributes file that is optimized for the dataset. This step is critically important for the correct calculation of tolerance and functional group metrics. The attributes file distributed with IDAS must be optimized for the dataset that is being processed. Optimization involves selecting the regional tolerance data that are appropriate for the dataset (for example, Southeast or Mid Atlantic) and allowing IDAS to select appropriate matches for tolerance and functional group data within the taxonomic hierarchy. This is most commonly done using the functions Create a new attribute file and Extract taxa from an abundance file.

7. **Calculate Community Metrics:** Once the attribute file has been optimized for the dataset, the Calculate Community Metrics module can be used to calculate taxonomic, tolerance, and functional group metrics.

8. **Calculate Similarities and Diversities:** The Calculate Diversities and Similarities module can be used to calculate various similarity and diversity indicies (Appendix I).

9. **Export data for use in reports or other software packages:** The Export Data module function Tab-delimited ASCII, full format taxa list for publication can be used to create a site-by-taxa table of the invertebrate data for inclusion in reports or posting as an electronic data table. The other Export Data module functions can be used to export data to other word processing, graphics, and statistical packages.

# Edit Data Module

The Edit Data module provides the user with a variety of options for editing, importing, and maintaining data and attribute files (table 10). The Edit data files option allows the user to subset data, combine and (or) delete data, summarize the distribution of taxa among sites and samples, identify ambiguous taxa, remove ambiguous parents, and resolve conditional/provisional data (fig. 11). The Import data option allows the user to convert non-NAWQA data to Bio-TDB format. The Maintain attributes file option is used to update or create attribute files that are specific to the taxa in a data file (abundance or taxonomic hierarchy) and (or) incorporate tolerance data from a specific region of the country. The Utilities option allows the user to convert a data matrix (taxa by sites or sites by taxa) to the stacked column format used by the Import data function and to randomly subsample a dataset. The Edit data module is the only module in the IDAS program that can use data in Bio-TDB, WR-EMAP, user-defined, and processed formats. The View, Exit, and About menus work the same in this module as in other modules. However, the Files menu in the Edit Data module differs from the files menus in other modules by having a series of submenu items (fig. 11). These submenus activate processing options that have different requirements for when and how they open files, spreadsheets, and (or) data tables.

When generating new files using the Edit Data module, IDAS will warn the user if the data produced by the program exceeds the data storage limits of Excel® or Access®. These limits vary depending on the amount of available memory and the version of the software used. Excel 2003 can store a maximum of about 65,198 rows of data whereas Excel 2007 can store as many as 1,048,576 rows of data. The number of rows of data that can be stored in Access 2003 and 2007 is limited only by the 2 gigabyte size limit for Access databases. In general, Access databases are preferred because they can hold larger datasets than can either version of Excel, they enforce strict data typing whereas Excel does not, and they offer better performance with IDAS than do Excel files.

## Subset Data

Selecting the Edit data files and Subset data menus brings up two additional submenus (by Sample information and by Taxonomy) that can be used to separate datasets into subsets based on sample identifiers (for example, SUID, STAID, CollectionDate, and sample type) or taxonomic groupings (for example, Trichoptera or Ephemeroptera). Selecting one of these submenus will call up the standard file selection (fig. 5) and data table or spreadsheet selection (fig. 6) window. However, since the Edit Data module can process files from both Bio-TDB and processed formats, the user must specify the type of file to subset by using the Select type of file to subset window (fig. 12) before the data table or spreadsheet selection window is displayed. Once the user selects a spreadsheet or data table to work on and the IDAS program establishes that the format is correct, the View menu is activated. The View menu can be used to display the first 20 lines of data.

**Table 10.** Options for editing, importing, and maintaining data and attribute files in the Edit Data module.

| Menu and submenus | Action |
|---|---|
| **Edit data files** | |
| Subset data | |
| by Sample information | Copies subsets of data into new Access data tables or Excel spreadsheets based on sample information (sites, collection dates, etc.) |
| by Taxonomy | Copies subsets of data into new Access data tables or Excel spreadsheets based on taxonomic information |
| Combine/delete data | |
| Combine tables/spreadsheets | Combines Access data tables or Excel spreadsheets of the same format (raw or processed) into a new data table or spreadsheet |
| Delete tables/spreadsheets | Deletes Access data tables or Excel spreadsheets |
| Summarize taxa | Summarizes the distribution of taxa among sites and samples and identifies ambiguous parents and children |
| Remove ambiguous parents | Allows the user to resolve ambiguous parent/child pairs outside of the Data Preparation module |
| Resolve conditional/provisional taxa | Allows the user to resolve conditional/provisional taxa outside of the Data Preparation module |
| **Import data** | |
| WR-EMAP format | Imports data in WR-EMAP format |
| User-defined formats | Imports data based on formats defined by the user |
| **Maintain attributes file** | |
| Create a new attribute file | |
| Extract taxa from an abundance file | Creates a new attributes file by extracting taxonomic data from an abundance file |
| Extract taxa from HIER in attribute file | Creates a new attributes file by extracting taxonomic data from a HIER file of an existing attributes file |
| Modify an existing file | |
| Add taxa from an abundance file | Adds taxa to an existing attributes file based on the contents of an abundance file |
| Refresh an EQTX spreadsheet | Refreshes the EQTXFG and EQTXTOL values in the EQTX worksheet of an existing attributes file |
| Update the ATTRIB spreadsheet | Update the tolerance or functional group data in the ATTRIB spreadsheet with new data |
| **Utilities** | |
| Convert matrix to columns | Converts a sites-by-taxa or taxa-by-site matrix to a stacked column format |
| Random subsample | Randomly selects a subsample based on counts |

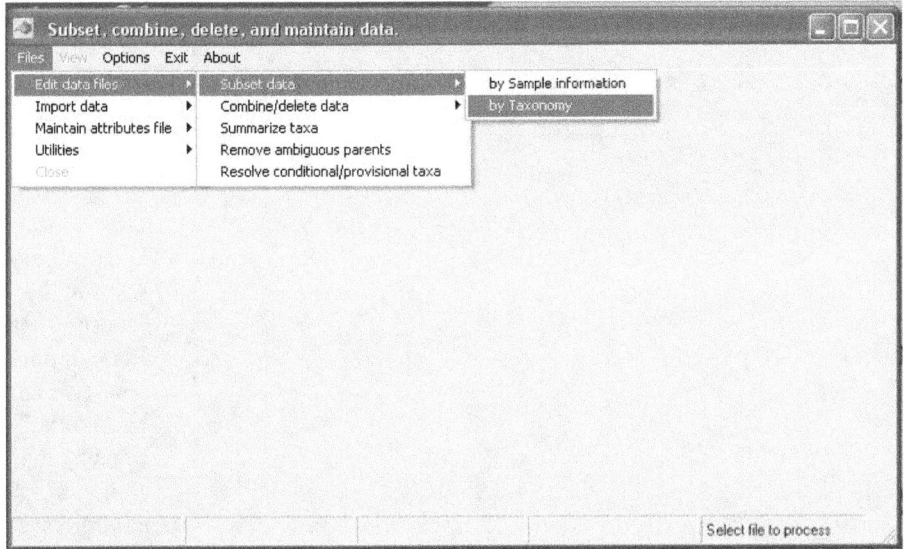

**Figure 11.** The opening window of the Edit Data module showing options for editing data files, importing data, maintaining attributes file, and utilities.

**Figure 12.** File-type selection window used in the Edit Data module.

[This example shows that the user has elected to process data that is stored in Bio-TDB format]

## Sample Information

The by Sample information option calls up a selection window (fig. 13) that allows the user to subset data based on any combination of SUID, STAID, Reach, Collection date, SampleID, SMCOD (sample code), and Sample Type (for example, R = RTH, D = DTH, Q = QMH). Once the user has selected an element of sample information to subset, the Sample Information check box is disabled (dimmed) until the user has indicated what action is to be applied to these data. The selected sample information is transferred to the Source file list box. Data to be subset are transferred between the Source file list box and the Destination file list box by double clicking the desired item or by highlighting the desired items (multiple selections can be made by using the shift and control keys in conjunction with the mouse) and clicking on the Transfer (>,<) buttons. The Accept button is used to prompt IDAS to act upon the user's selections. The selections

in figure 13 change when WR-EMAP data are processed. "Reach" changes to "Visit number" and the selections for "Sample Type" change to "R," "S," and "T" to correspond to EMAP sample types "Reachwide," "Shore," and "Targeted Riffle," respectively.

Once the user selects the Accept button, the appropriate Sample Information check box is checked, provided that the user actually selected items to transfer to the destination file. If the user did not select items to send to the destination file, then the Sample Information check box will remain unchecked. In this way, the user can quickly see what items of sample information have been selected. The user can click on previously selected Sample Information check boxes to review or modify selections. If one or more boxes are checked, the Process button will be activated, which implements the subsetting procedures. Once the Process button has been activated and the user has finished selecting items to subset by, subsetting is implemented by clicking on the Process button. The user will be prompted for the name of a new spreadsheet or data table in which to store the subset of data (fig. 14) or warned that the combination of sample information selected did not correspond to any data in the source file. The name of the new spreadsheet or data table must be 15 characters or less and contain only letters, numbers, and the underscore character. The IDAS program will warn the user if the name entered does not meet these criteria. The new spreadsheet or data table is stored in the original workbook or database. The IDAS program will display **FINISHED** in the right-hand status bar panel (fig. 13) when it has finished processing and saving data. The Files and Close menus can then be used to reset the module and begin processing another dataset.

The by Sample information subsetting option is a very important feature of the IDAS program. It allows the analyst to select samples that need to be processed together. For

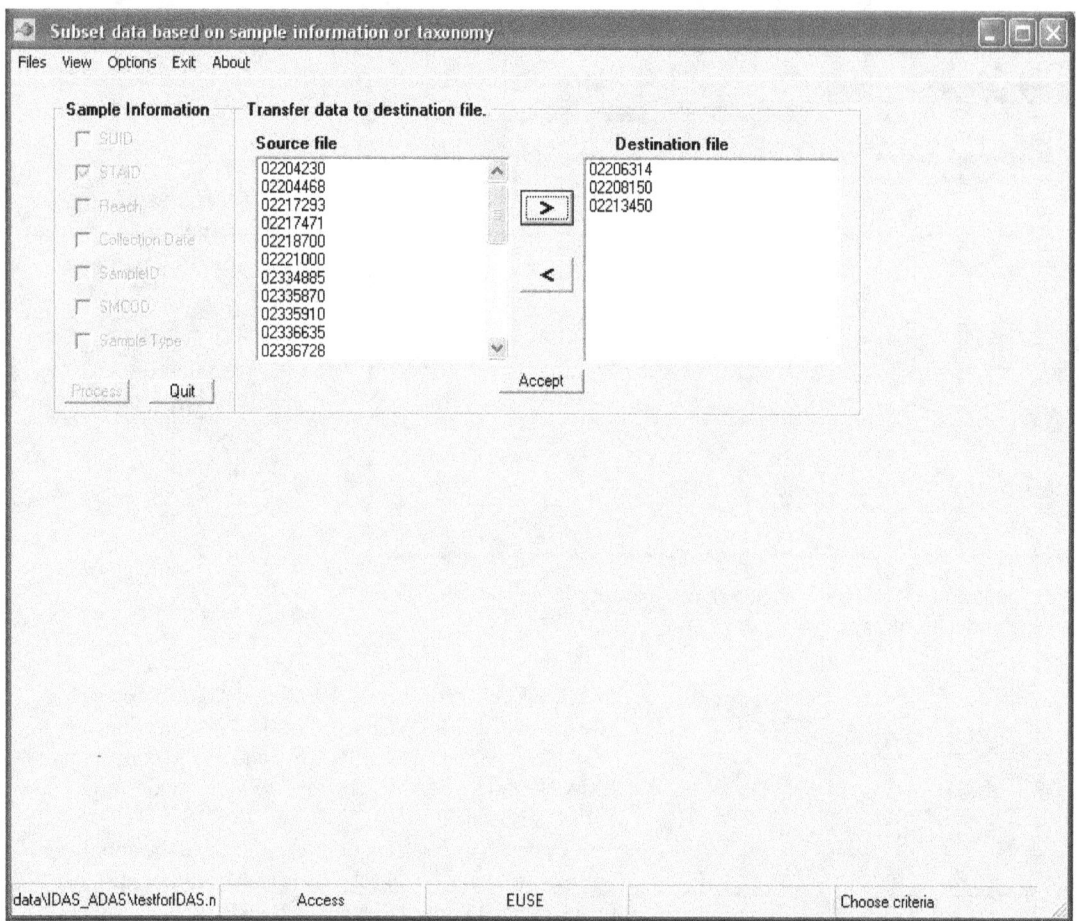

**Figure 13.**    Selection window used to subset data based on sample information.

**Figure 14.**    The Enter a table name to store data window is used to enter a table or spreadsheet name in which to store the data.

[This window displays the data tables or spreadsheets that currently exist in the workbook. The IDAS program checks to make sure that the table or spreadsheet name that is entered meets the criteria listed in this window]

example, the analyst can create datasets that consist only of trends sites, or sites that are part of specific topical studies (for example, land-use gradient studies), or data from specific time periods. This subsetting option also can be used to separate qualitative samples which should be analyzed separately, from quantitative samples. Breaking down the large datasets exported from Bio-TDB into smaller, more coherent datasets is important for data analysis and program performance because smaller files can be processed faster and more efficiently than larger ones.

## Taxonomy

The by taxonomy option calls up a new selection window (fig. 15) that allows the user to select subsets of data based on taxonomy. Taxonomic groups are selected by clicking on the appropriate check boxes, which are activated only if the taxonomic group exists in the dataset. Selected taxa are sent to a new spreadsheet or data table by clicking the Process button. The user will be prompted for a valid spreadsheet or data table name (fig. 14), and the new data will be stored in the original Excel workbook or Access database. After saving the selected data, the IDAS program updates the

selection window by deactivating (dimming) the check boxes that were previously selected. That is, active check boxes correspond only to taxa that exist in the dataset and that have not yet been saved to a new spreadsheet or data table. In this way, a dataset can be iteratively subset until no taxa remain and only the Quit button remains active. The Quit button or the menu commands Files/Close can be used to reset the subset by taxonomy function and return to the opening screen of the Edit Data module.

Subsetting by Taxonomy can be used to create separate datasets for specific taxonomic groups. These datasets can then be processed through the other modules to calculate metrics or apply different data preparation options to specific taxonomic groups. For example, the analyst may want to process midges (Chironomidae) with a different lowest taxonomic level than other invertebrates. Alternatively, the analyst may want to preserve information on the lifestages of beetles (Coleoptera) but not other invertebrates. Once the subsets of data have been created and processed, they can be recombined by using the Combine/delete option of this module and run through other IDAS modules to calculate metrics, diversities, similarities, or to export data to other analysis packages.

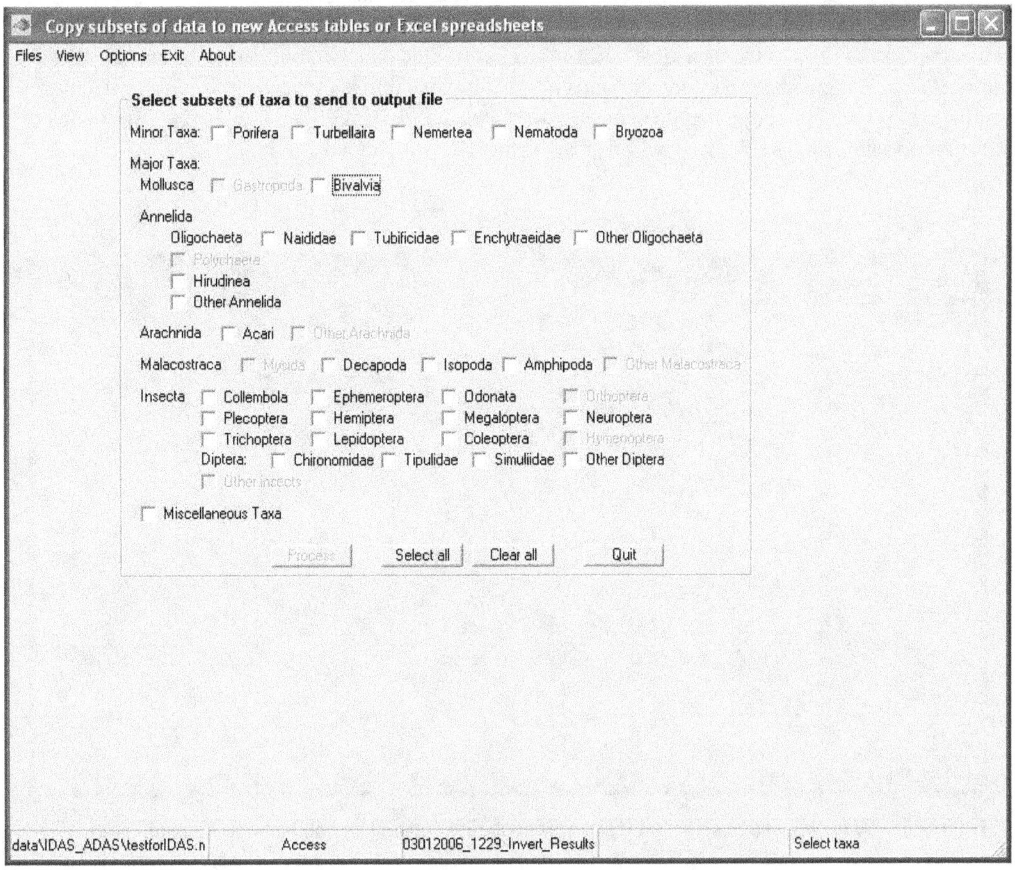

**Figure 15.** Selection window used to subset data based on taxonomic information.

[Taxonomic groups that are not present in the source data file are disabled (dimmed)]

## Combine/Delete Data

The Combine/delete data menu has two submenus (Combine tables/spreadsheets and Delete tables/spreadsheets) that allow the user to form new data tables or spreadsheets by combining Excel spreadsheets or Access tables of similar types (that is, either Bio-TDB or processed format) or to delete spreadsheets or data tables of all types. Selecting either option calls up a standard file-selection window (fig. 5) that is used to open the workbook or database containing the spreadsheets or data tables. If the Combine tables/spreadsheets option is selected, the user must select the type of file to combine by using the Select type window (fig. 12). This window will not appear if the Delete tables/spreadsheets option is selected, as this option applies to any spreadsheet or data table. Once the user selects a workbook or database, the IDAS program lists the spreadsheets or data tables of the selected type in the Source table list box and activates the View menu (fig. 16).

Both the **combine** and **delete** functions use a **source** (left) and **destination** (right) list box to select tables or spreadsheets to combine or delete (fig. 16). The captions for the source and destination boxes, command buttons, and frames will change depending upon whether data are being combined or deleted. The contents of a spreadsheet or data table highlighted in either list box can be viewed by using the View menu item, which will display the first 20 lines of the data table. Data tables or spreadsheets can be transferred between the **source** and **destination** list boxes by highlighting the file name(s) and selecting the appropriate Transfer (< or >) button or by

> **WARNING: Once data tables or spreadsheets have been deleted in the Edit Data module, they cannot be recovered! Exercise caution when using the delete capability.**

double-clicking the data table or spreadsheet file name. Data are combined or deleted by selecting the Combine or Delete buttons. The IDAS program prompts the user for the name of a new data table or spreadsheet in which to store the combined data (fig. 14), but it gives no further warning if data tables or spreadsheets are being deleted. Once data tables or spreadsheets have been deleted, they cannot be recovered. Therefore, caution should be exercised when selecting the delete option. When data tables or spreadsheets are combined, the original data tables or spreadsheets are kept intact. The **combine** and **delete** functions use the Exit option on the menu to reset the module. This returns the user to the opening screen.

The **combine** function can be very useful for recombining datasets that were subset on the basis of taxonomy or sample information. The recombined dataset can then be run through the other IDAS modules to calculate community metrics, diversities, or similarities, or exported to other software packages. The **delete** function allows the user to eliminate unwanted data without having to exit the IDAS program. This function is particularly useful for deleting Access tables, because Access does not allow the user to delete multiple tables simultaneously.

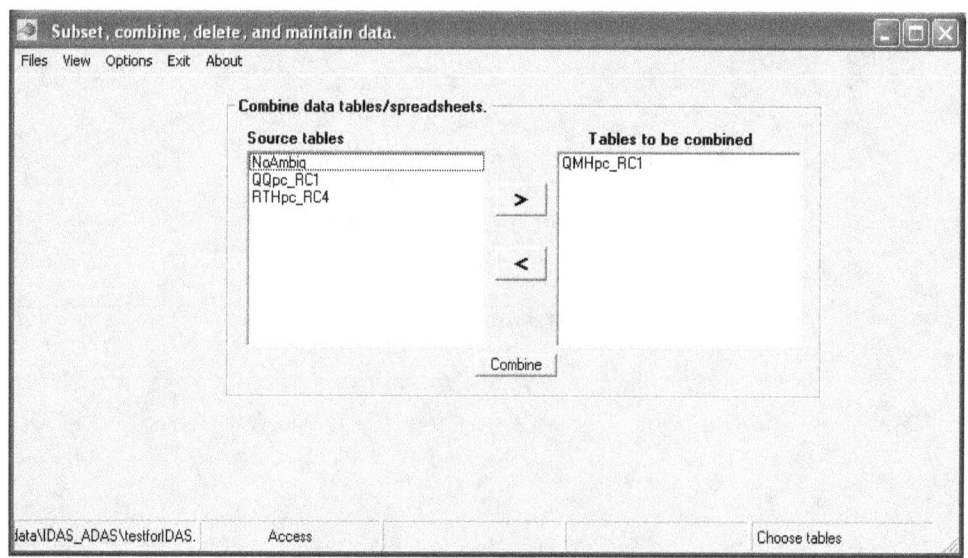

**Figure 16.**   Window used to select tables or spreadsheets that will be combined or deleted.

[The combine option is shown in this figure. The left-hand list box displays the source tables and the right-hand list box displays the tables or spreadsheets that will be deleted or combined. The captions of the frame and list boxes change depending upon the processing option that has been selected]

## Summarize Taxa

The Summarize taxa menu is used to identify ambiguous taxa and calculate statistics on the distribution of taxa among samples and sites. Selecting this option from the Files menu calls up a file-selection window (fig. 5), followed by a file-type selection window (fig. 12), and then a data table or spreadsheet selection window (fig. 6). The file-type selection window is required because the Summarize taxa option can be used to process data in both Bio-TDB (table 1) and processed formats (table 3). The user also has the option of summarizing distributions based on BU_ID or the combination of BU_ID and lifestage using the Use lifestage in summarization window (fig. 17). Once the user selects a spreadsheet or data table to work on and the IDAS program establishes that the format is correct, the View menu item is activated. The View menu is used to view the first 20 rows of the selected spreadsheet or data table.

Results are stored in the source file using a data table (Access) or spreadsheet (Excel) name that contains the input data table name (for example, RTH) plus the suffix "_Distrib" (for example, RTH_Distrib). The Summarize taxa function identifies ambiguous taxa and calculates distributions across the entire contents of the input data table or spreadsheet. The summary contains the list of taxa (rows) in phylogenetic order with the taxonomic hierarchy plus the following columns of information:

**BU_ID:** name of the taxon being summarized.

**Lifestage:** "A" for adult, "P" for pupae, "L" for larvae, or blank.

**Ambig:** indicates if the taxon is ambiguous ("Yes") or not (blank).

**noChild:** number of children that are associated with an ambiguous taxon.

**sumChild:** sum of the abundances of children associated with an ambiguous taxon.

**pSumChild:** percentage of total abundance represented by sumChild.

**Sites:** number of sites where the taxon occurs.

**pSites:** percentage of sites where the taxon occurs.

**Samples:** number of samples in which the taxon occurs.

**pSamples:** percentage of samples in which the taxon occurs.

**Abund:** the abundance of the taxon summed in all samples.

**pAbund:** percentage of total abundance represented by Abund.

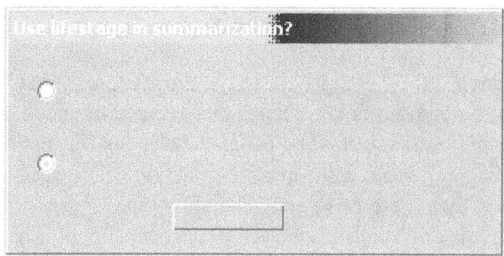

**Figure 17.** The distribution of taxa can be summarized on the basis of BU_ID or the combination of BU_ID and lifestage.

**Ave_All:** average abundance of taxon in all samples.

**MAX_All:** maximum abundance of taxon in all samples.

**MIN_All:** minimum abundance of taxon in all samples.

**StDev_All:** standard deviation of abundance of taxon in all samples.

**Ave_Occur:** average abundance of taxon in samples where it occurs.

**MAX_Occur:** maximum abundance of taxon in samples where it occurs.

**MIN_Occur:** minimum abundance of taxon in samples where it occurs.

**StDev_Occur:** standard deviation of abundance of taxon in samples where it occurs.

Average, maximum, minimum, and standard deviation of abundances are given based on all (_ALL) samples (that is, abundance is considered to be zero for samples in which the taxon does not occur) and based only on samples in which the taxon actually occurs (_Occur). Processing data through the Summarize taxa function is an important preliminary step in the analysis of invertebrate data. The statistics generated by this function allow the analyst to get a comprehensive overview of the data, which can be helpful in deciding how to resolve ambiguities, subset data, set limits for deleting rare taxa, or in providing a taxa list for a group of samples.

> **TIP: Use the "Summarize taxa" function of the Edit Data module to investigate how to resolve ambiguous taxa and to set limits for deleting rare taxa.**

## Remove Ambiguous Parents

The Remove ambiguous parents option is used to remove ambiguous taxa (parents) that occur at or above a taxonomic level specified by the user. Selecting this option from the Files menu calls up a file-selection window (fig. 5), followed by a file-type selection window (fig. 12), and then a data table or spreadsheet selection window (fig. 6). The file-type selection window is required because the Remove ambiguous parents option applies to files in both Bio-TDB (table 1) and processed formats (table 3). The Remove ambiguous taxa frame (fig. 18) will be displayed once the data table or spreadsheet has been opened. The Select taxa level frame allows the user to select the taxonomic level (genus, tribe, subfamily, family, suborder, order, or class) used to remove ambiguous parents. Taxa that are identified as ambiguous and that occur at or above the selected taxonomic level are removed from the dataset. The selections in the Identify ambiguous taxa frame determine the method used to identify ambiguous taxa within the dataset. The selection Separately for each sample identifies and removes ambiguous parents by considering each sample separately. The ambiguous taxa in a sample are identified and removed if they are at or above the selected taxonomic level, and the data are saved before processing data from the next sample. The selection For a group of samples combines all samples and

identifies ambiguous parents in the combined data. Ambiguous parents that are at or above the selected taxonomic level are then removed from the individual samples. The selections in the Options for combining lifestages frame allows the user to identify ambiguous taxa by taking lifestage information into account (Keep lifestages separate) or by ignoring the lifestage information (Combine lifestages). Once the processing options have been selected, the Process button becomes activated and can be used to remove ambiguous taxa. The Quit button resets the module and returns the user to the opening screen of the Edit Data module without processing any data.

## Resolve Conditional and Provisional Taxa

The Resolve conditional and provisional taxa option is used to resolve conditional and provisional taxa that occur in the dataset. Conditional and provisional identifications occur when the NWQL BG cannot definitively identify an organism but can determine that it is similar to another taxon (nr.), an undescribed species or genus (sp. 1, genus A), or part of a group of indistinguishable taxa (group, complex, sp. 1/sp. 2). Provisional and conditional taxa are indicated by the appearance of "nr.," "cf.," "/," "group," "complex," "n. sp.," or "?" in the taxonomic designation or by a species or genus name that consists of a single letter or number (sp. 1, sp. 2, or genus A; see Moulton and others, 2000). Provisional and conditional taxa provide information on taxonomic affinities that would be lost if identifications were reported only at higher definitive taxonomic levels (for example, genus or family). Selecting this option from the Files menu calls up a file-selection window (fig. 5), a file-type selection window (fig. 12), and a data table or spreadsheet selection window (fig. 6). The file-type selection window is required because the Resolve conditional and provisional option applies to files in both Bio-TDB (table 1) and processed formats (table 3).

The Resolve conditional and provisional taxa window (fig. 19) provides tools for viewing conditional and provisional taxa and resolving them by assigning their abundances to other taxa. A counter (Reviewing taxon ❶) lists the total number of provisional/conditional taxa and tracks progress through the list. The Original BU_ID field (❷) contains the name of the conditional/provisional taxon as it occurs in the dataset. The Name field (❸) contains the taxon name that is being considered as a replacement for the provisional/conditional taxon. This field is initially populated with

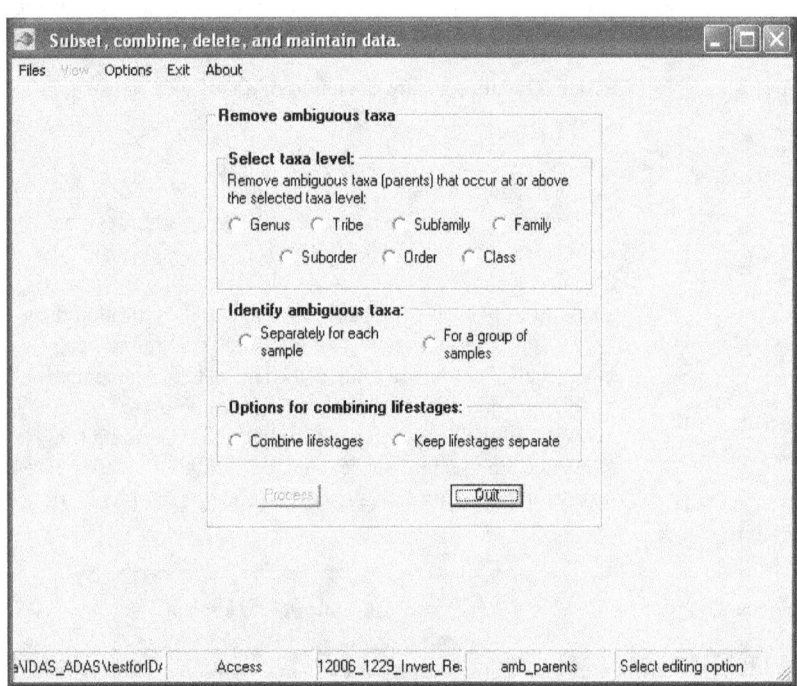

**Figure 18.**    The Remove ambiguous taxa function can be used to remove ambiguous taxa that occur at or above a user-specified taxa level.

[Ambiguous taxa can be identified separately for each sample or for the entire dataset. The user may also elect to consider lifestage information when identifying ambiguous taxa]

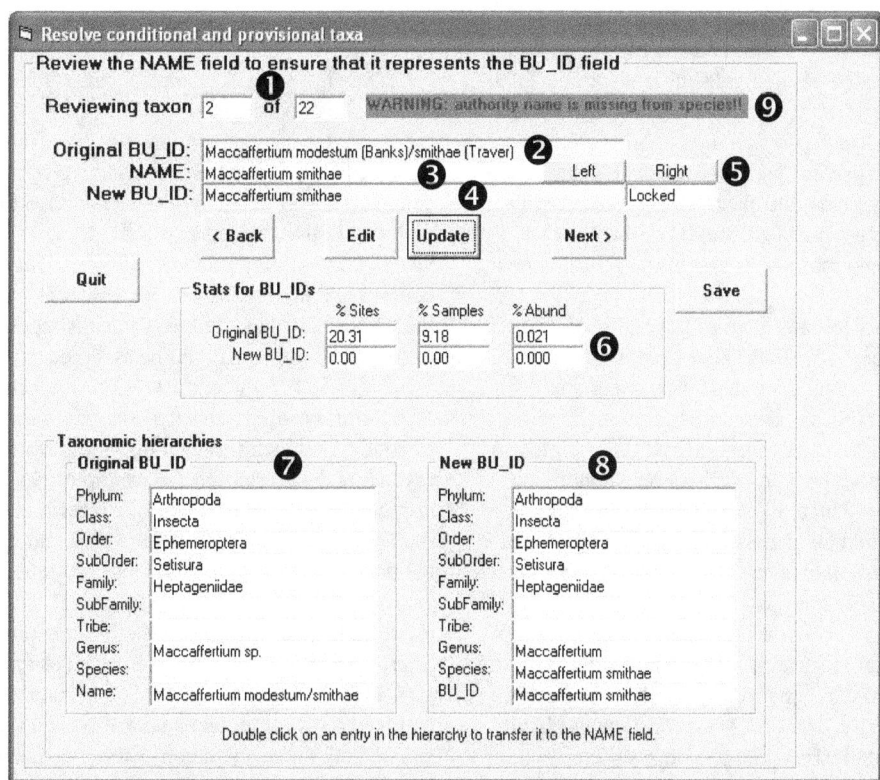

**Figure 19.** Resolve conditional and provisional taxa window allows the user to assign data associated with provisional/conditional taxa to other taxa.

[❶ counter showing position in the lost of provisional/conditional taxa; ❷ provisional/conditional taxon as it appears in the data; ❸ name to be substituted for the provisional/conditional name as suggested by the program or entered by the user; ❹ name that will be substituted for the provisional/conditional name when the Update button is selected; ❺ conversion assistance buttons that automatically suggest names for the provisional/conditional taxon; ❻ percentage of sites, samples, and total abundance associated with the Original BU_ID and New BU_ID names; ❼ taxonomic hierarchy associated with the Original BU_ID; ❽ taxonomic hierarchy associated with the New BU_ID; ❾ warning message indicating that the New BU_ID name lacks an authority name]

the Original BU_ID name without the authority name(s). The New BU_ID field (❹) contains the name that will be used to replace the Original BU_ID name. The color of these three text fields indicates whether the text box can be edited (white) or not (yellow). The conversion assistance buttons (❺) are used to populate the Name field by extracting the taxon name(s) from the provisional/conditional taxon name (Original BU_ID). The Name field can be manually edited so the user can modify the name suggested by the conversion assistance buttons or enter a completely different name. The Back and Next buttons are used to move through the list of provisional/conditional taxa. The Update button is used to substitute the New BU_ID name for the Original BU_ID name in the dataset. If the user moves to another provisional/conditional taxon without clicking on the Update button, the Original BU_ID will not be replaced with the New BU_ID. Once the taxon has been updated by clicking on the

Update button, the Locked field is displayed and clicking of the conversion assistance button will no longer affect the New BU_ID field. The New BU_ID field can be unlocked by double clicking on the Locked field until it disappears. Changes made with the Update button can be modified until the results are saved to a new data file using the Save button. The Quit button is used to exit this option, reset the Edit Data module, and return to the opening screen without making any changes to the data.

The Stats for BU_IDs fields (❻) display the percentage of sites, samples, and total abundance in the data that are associated with the Original BU_ID and New BU_ID names. This information can be used to help decide whether or not to assign a conditional/provisional taxon to another taxon. For example, if *Maccaffertium modestum* exists in the dataset but *Maccaffertium smithae* does not, then the user might want to assign the abundance of *Maccaffertium modestum/smithae* to

*Maccaffertium modestum.* The **Taxonomic hierarchies** associated with the **Original BU_ID** (❼) and **New BU_ID** (❽) also can be used to update the **Name** field simply by double clicking on the name in the taxonomic hierarchy. For example, if the user decides that *Maccaffertium modestum/smithae* should be reported as *Maccaffertium* sp. or Heptageniidae, this can be accomplished simply by double clicking on *Maccaffertium* or Heptageniidae in the taxonomic hierarchy of the **New BU_ID** (note that sp. should not be used as part of the genus name in the **Name** field).

The conversion assistance buttons (❺) automate the process of populating the **Name** field. They do this by removing the provisional/conditional identifier (for example, nr., /, sp. 1, cf.) and taxonomic authority names, when present, from the **Original BU_ID** name. The conversion assistance button(s) change depending upon the type of provisional/conditional taxon that is being evaluated. Figure 19 shows an example where a taxon could not be distinguished between two species, *Maccaffertium modestum* (Traver) and *Maccaffertium smithae* (Banks). Clicking on the left conversion assistance button (**Left**) places *Maccaffertium modestum* in the **Name** field (❸) and causes the program to search for *Maccaffertium modestum* in the data file and populate the **New BU_ID** field (❹) with the name as found in the data file (*Maccaffertium modestum* (Banks)). Clicking on the right conversion assistance button (**Right**) populates the **Name** field with *Maccaffertium smithae* and searches for a match in the data file. *Maccaffertium smithae* does not exist in the data file, so the **New BU_ID** field is populated with *Maccaffertium smithae* without an authority name, a warning message is displayed (❾), and the **Edit** button appears. The **Edit** button is used to manually modify the **New BU_ID** field. Clicking on the **Edit** button locks the **Name** field (yellow) and opens the **New BU_ID** field for editing (white). This is most commonly used to add an authority name to a species that does not already occur in the dataset, for example, change *Maccaffertium smithae* to *Maccaffertium smithae* (Traver).

Clicking on the **Save** button substitutes the **New BU_ID** names for the **Original BU_ID** names as specified by the **Update** button. The "save data as" window (fig. 14) is displayed and the data are stored as a new spreadsheet or data table in the original Excel or Access file.

## Import Data Files

The Import data files option automates the process of converting non-NAWQA data into Bio-TDB format. Non-NAWQA data can be manually converted to Bio-TDB format, but experience has shown that this process is prone to error particularly when creating data files in Excel, which does not enforce strict data typing. Consequently, the functions in the Import data files and Utilities options should be used to convert data to Bio-TDB format.

## WR-EMAP Format

Data in WR-EMAP format data (table 11, WR_EMAP.xls) can be imported using the WR-EMAP format option in the Import data file function (fig. 20). The WR-EMAP format was used early in the USEPA EMAP program to store data collected from sites in the Western United States. It has been included in IDAS because it provides an alternative mechanism for identifying provisional/conditional taxa. The **Distinct** column in the WR-EMAP data format is used to flag taxa that the taxonomists have judged to be distinct from one another even though they could not be assigned non-ambiguous taxa names (table 12). In contrast, NAWQA Program (Bio-TDB) taxonomists distinguish such taxa by giving them a provisional or conditional identifier that provides information on the taxon that they resemble, for example *Hydropsyche* sp. nr. *elissoma* Ross. The IDAS program recognizes that taxa with the **Distinct** flag set to "Yes" are different from other taxa in the dataset by appending "_D" and a sequential number to the taxon name, for example *Baetis* sp._D1 and *Baetis* sp._D2 (table 12). Taxa with the distinct designation (*Baetis* sp._D1, *Baetis* sp._D2) will not be interpreted as ambiguous taxa by IDAS whereas taxa without this designation (*Baetis* sp.) would be identified as ambiguous taxa. Taxa with the same names can be both distinct and non-distinct taxa in the same sample (*Baetis*, table 12).

Importing data in WR-EMAP format begins by selecting a file (fig. 5) and spreadsheet or table to process (fig. 6) and then specifying a table/spreadsheet in which to store the processed data (fig. 14). Variables in the WR-EMAP format are converted to variables in Bio-TDB format according to the relations defined in table 13. SampleIDs are created by combining the WR-EMAP variables SITE_ID, VISIT_NO, DATE_COL, SAMP_ID, and SAMPTYPE (table 11) to form a unique sample identifier and then assigning sequential numbers to each sample identifier. IDAS prompts the user (fig. 21) to provide a starting number for the SampleIDs and an interval between successive SampleIDs. The variable SMCOD is formed by using "EMAP" for the SUID + 2-digit representations of the month and year from DATE_COL (0799) + "I" (invertebrate sample) + the sample type identifier (SAMPTYPE: R = reachwide, S = shore, T = targeted riffle) + "M" (main body sample) + a 4-digit sequential number for each SampleID (EMAP0799ISM0010). SortCodes are created

> **TIP:** The use of "EMAP" as a SUID is restricted to data imported as WR-EMAP data. "EMAP" should not be used as a SUID for data imported into IDAS by other means.

**Table 11.**    Structure of the WR-EMAP invertebrate data file (spreadsheet Data in WR_EMAP.xls).

[reference (ref.)]

| Column name | Data type | Example | Comments |
|---|---|---|---|
| TAXANAME | Text | AMELETUS SP. | Taxon name |
| SITE_ID | Text | WCAP99-0503 | Site identifier |
| YEAR | Integer | 2000 | Year sample was collected |
| VISIT_NO | Integer | 1 | Visit number for the year |
| SAMPTYPE | Text | REACHWIDE | Sample type |
| SAMPLED | Text | Yes | Site sampled? |
| DATE_COL | Date/time | 10/3/2000 | Date sample was collected |
| TEAM_ID | Integer | 1 | Time identification number |
| SAMP_ID | Integer | 260930 | Sample identifier |
| LAB_NAME | Text | ABC Labs | Name of laboratory that processed the sample |
| EA_TAXON | Integer | | Laboratory code |
| ABL_CODE | Integer | 55 | Laboratory code |
| ABUND | Double | 2 | Abundance for taxon in sample |
| LARVAE | Double | 2 | Number of larvae |
| PUPAE | Double | 0 | Number of pupae |
| ADULTS | Double | 0 | Number of adults |
| PHYLUM | Text | ARTHROPODA | Phylum |
| CLASS | Text | INSECTA | Class |
| ORDER | Text | EPHEMEROPTERA | Order |
| FAMILY | Text | AMELETIDAE | Family |
| SUBFAM | Text | NA | Subfamily |
| TRIBE | Text | NA | Tribe |
| GENUS | Text | AMELETUS | Genus |
| SPECIES | Text | NA | Species |
| DISTINCT | Text | YES | Distinct taxon |
| LR_TAXA | Text | NO | Large or rare taxon |
| ITISNUM | Integer | 100996 | Taxa identification number |
| COM_LAB | Text | immature | Laboratory comment field |
| COM_IM | Text | ref. | Comment field |

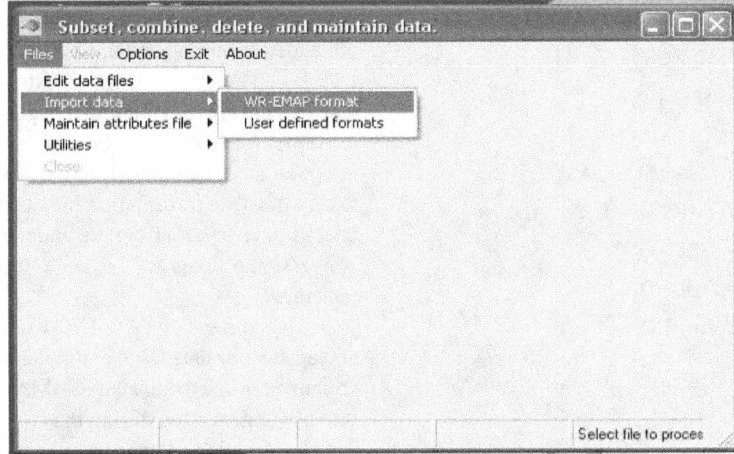

**Figure 20.**    The Import data option can import data in WR-EMAP or user-defined formats.

**Table 12.**  The Distinct column in WR-EMAP data files is used to identify taxa that are distinct from other taxa even though a unique name could not be assigned to them (*Baetis* sp., *Agabus* sp.).

[IDAS recognizes distinct taxa by appending "_D" and a sequential number to the taxon name. L, larva; A, adult]

| Site | WR-EMAP name | Lifestage | Distinct | BU_ID name |
|------|--------------|-----------|----------|------------|
| ZAP-1 | *Baetis* sp. | L | No | *Baetis* sp. |
| ZAP-1 | *Baetis* sp. | L | Yes | *Baetis* sp._D1 |
| ZAP-1 | *Baetis* sp. | L | Yes | *Baetis* sp._D2 |
| ZAP-1 | *Baetis tircaudatus* | L | Yes | *Baetis tricaudatus* |
| ZAP-1 | *Agabus* sp. | L | Yes | *Agabus* sp._D1 |
| ZAP-1 | *Agabus* sp. | A | Yes | *Agabus* sp._D2 |

**Table 13.**  Variables in the WR-EMAP file format are converted to Bio-TDB variables using the following correspondence between variables.

[An empty field indicates that the variable is not used in the WR-EMAP dataset]

| Bio-TDB variable | WR-EMAP variable |
|------------------|------------------|
| SampleID | Created by IDAS |
| SMCOD | Created by IDAS |
| STAID | SITE_ID |
| Reach | VIST_NO |
| CollectionDate | DATE_COL |
| Phylum | PHYLUM |
| Class | CLASS |
| Order | ORDER |
| SubOrder | |
| Family | FAMILY |
| SubFamily | SUBFAM |
| Tribe | TRIBE |
| Genus | GENUS |
| Species | SPECIES |
| BU_ID | TAXANAME |
| SortCode | Created by IDAS |
| Lifestage | LARVAE, PUPAE, ADULT |
| Notes | COM_LAB |
| LabCount | ABUND |
| Abundance | ABUND |

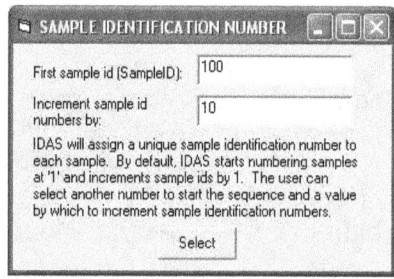

**Figure 21.**  The starting number and interval between consecutive SampleIDs can be specified by the user when importing WR-EMAP data.

by extracting the taxa (BU_ID) and associated taxonomic hierarchy, sorting this list using the criteria specified in table 7, and then numbering the sorted taxa. Lifestage information is based on the information stored in the PUPAE and ADULT columns of the WR-EMAP file. The numbers of larvae are determined by subtracting the number of pupae and adults from the total abundance (ABUND). Separate lines of data are added for each lifestage associated with the taxon. The COM_LAB column is used to extract sample processing notes (Notes) that correspond to notes generated by the NWQL, for example, "immature." LabCount and Abundance columns are populated with data from the ABUND column after taking lifestage information into account. The IDAS program checks to ensure that sample abundances before and after the conversion process are the same. If not, an error message is generated.

The conversion of WR-EMAP files creates a second spreadsheet or data table that holds the sample information contained in the original WR-EMAP data file and the SampleID generated during the conversion process (table 14). This information is stored in a spreadsheet or table that is named by adding the suffix "_Samp_Info" to the name used

**Table 14.** Sources (WR-EMAP) for data that are stored in the sample information table (Samp Information) created when WR-EMAP data are imported.

| Samp Information | Source in WR-EMAP |
|---|---|
| SampleID | Created by IDAS |
| STAID | SITE_ID |
| Year | YEAR |
| REACH | VIST_NO |
| SAMPTYPE | SAMPTYPE |
| SAMPLED | 1 = YES |
| COLLECTIONDATE | DATE_COL |
| TEAM_ID | TEAM_ID |
| SAMP_ID | SAMP_ID |
| LAB_NAME | LAB_NAME |

to store the abundance data (for example, EMAP_Samp_Info in WR_EMAP.xls). The purpose of the sample information file is to maintain a tight association between the SampleID created by IDAS and the original WR-EMAP sample and site identifiers.

If the user intends to convert the WR-EMAP data to densities (no./m²), then the equivalent of a "_Sample_All" spreadsheet or data table has to be created (for example, EMAP_Sample_Areas.xls). The suggested format for this file is given in table 15. This format provides detailed information on the original WR-EMAP sample identifiers as well as the IDAS identifiers (for example, SampleID). The area sampled (AREASAMPTOT) needs to be in square centimeters (cm²) in order for IDAS to correctly calculate densities as number per square meter (no./m²).

## User-Defined Formats

The User defined formats option creates a Bio-TDB formatted file (table 1) based on user supplied files (Excel or Access 2003 or 2007) that contain information on taxa abundances, taxonomic hierarchies, and sample and site identifiers. This information may be contained in a single file or distributed among multiple files, but all files must be in stacked column rather than matrix format. If data are in matrix format (for example, sample-by-taxa or taxa-by-sample matrices), then the Convert matrix to columns function in the Utilities menu must be used to convert the matrix to stacked column format before importing the data. At a minimum, the data to be imported must include:

1. One or more sample identifiers that, taken together, uniquely identify each sample.

2. Abundance and taxonomic hierarchy for each taxon in the data. IDAS supports nine taxonomic levels: phylum, class, order, suborder, family, subfamily, tribe, genus, and species.

3. Date when each sample was collected. The data format must be compatible with Excel or Access.

4. A sample site identifier (text or numeric).

In addition, if the sample contains a mix of qualitative and quantitative samples, the data file must include one or more sample type identifiers (text or numeric) that distinguish quantitative samples from qualitative samples. If the data contain only qualitative or quantitative samples, then the sample type identifier is not needed.

The process of importing data using user-defined formats begins by selecting the User defined format option of the Import data menu. The program prompts the user to select

**Table 15.** Suggested format for creating an Excel spreadsheet or Access table that contains information on the area sampled for data that are in WR-EMAP format (EMAP_Sample_Areas.xls).

| Column name | Data type | Comments |
|---|---|---|
| SAMPLEID | Integer | IDAS-generated sample identifier |
| STAID | Text | Station identifier |
| YEAR | Integer | Collection year |
| REACH | Text | Visit number or sampling reach |
| SAMPLETYPE | Text | Sample type: RWS, SHS, TRS |
| SAMPLEMEDIUMCODE | Text | Invertebrate (I) |
| SAMPLED | Text | Site sampled (Yes = 1)? |
| COLLECTIONDATE | Date/time | Collection date |
| TEAM_ID | Integer | Team identifier |
| SAMP_ID | Integer | EMAP sample identifier |
| LAB_NAME | Text | Laboratory name |
| AREASAMPTOT | Double | Area sampled in square centimeters (cm²) |

three spreadsheets or tables that contain the abundance data, sample information, and taxonomic hierarchy. Standard file (fig. 5) and spreadsheet or table (fig. 6) selection windows are used to open each spreadsheet or table. The abundance data, sample information, and taxonomic hierarchy may be contained in a single spreadsheet/table (table 16, Import_ Format_Example_1.xls), which would be opened three times or distributed among two (table 17, Import_Format_ Example_2.xls; table 18, Import_Format_Example_3.xls) or three spreadsheets/tables (table 19, Import_Format_ Example_4.xls) in the same or different Excel or Access files (Appendix II).

The first spreadsheet or table that is opened must contain the abundance data, the name of the taxon, and one or more sample identifiers (numeric or alphanumeric) that, singly or in combination, uniquely identify each sample. The second spreadsheet/table (sample information) that is opened must contain the sample-collection date, station identifiers, and sample identifiers that, singly or in combination, uniquely identify each sample and match the sample identifiers in the abundance file. The third spreadsheet/table (taxonomic hierarchy) that is opened must contain a taxon identifier that matches each taxon in the abundance spreadsheet or table with the appropriate taxonomic hierarchy.

**Table 16.**    Example of the structure of a data file (Import_Format_Example_1.xls) in which the same spreadsheet (Abund_w_SampInfo_Hier) provides abundance, taxonomic, and sample information to the data import functions.

[no./m$^2$, number per square meter; gr., group; %, percentage; Rep, replicate; Subsamp, subsample]

| Column name | Example data | Data type | Comment |
|---|---|---|---|
| Site number | 02341744 | Text | Site identifier number |
| Stream name | Cat Creek | Text | Stream name |
| Sampler | Hess | Text | Sampler |
| Habitat | Riffle | Text | Habitat sampled |
| Site abbreviation | ARBQR | Text | Site abbreviation |
| Rep | 2 | Integer | Replicate sample number |
| Date | 37483 | Date | Sample collection date |
| Subsamp | 18.76 | Double | Subsample (%) |
| Year | 2002 | Integer | Year of sample collection |
| Phylum | Arthropoda | Text | Phylum |
| Class | Insecta | Text | Class |
| Order | Diptera | Text | Order |
| Suborder | Nematocera | Text | Suborder |
| Family | Chironomidae | Text | Family |
| SubFamily | Chironominae | Text | Subfamily |
| Tribe | Chironomini | Text | Tribe |
| Genus | *Microtendipes* | Text | Genus |
| Species | *Microtendipes pedellus* gr. | Text | Species |
| Taxon | *Microtendipes pedellus* gr. | Text | Taxon name |
| Density | 10.661 | Double | Density (no./m$^2$) |

**Table 17.** Example of the structure of a data file (Import_Format_Example_2.xls) in which the information needed to import the data is contained in two spreadsheets; one containing sample information and abundance data and a second containing the taxonomic hierarchy.

[Abund_w_SampInfo contains the sample information and abundance. Hier contains the taxonomic hierarchy. The data in the two spreadsheets are linked by the Taxon and Organism columns. %, percentage; gr., group; no./m², number per square meter]

| Spreadsheet | Column name | Example data | Data type | Comment |
|---|---|---|---|---|
| Abund_w_SampInfo | | | | |
| | Sample code | ARBQR02 | Text | Sample identifier number |
| | Station Number | 02341744 | Text | Site identifier number |
| | Stream | Cat Creek | Text | Stream name |
| | Sample Type | Hess | Text | Sampler |
| | Habitat | Riffle | Text | Habitat sampled |
| | Site abbreviaton | ARBQR | Text | Site abbreviation |
| | Replicate | 2 | Integer | Replicate sample number |
| | Sampling Date | 8/15/2002 | Date | Sample collection date |
| | Subsample | 18.76 | Double | Subsample (%) |
| | Year | 2002 | Integer | Year of sample collection |
| | Taxon | *Microtendipes pedellus* gr. | Text | Taxon name |
| | Density | 10.661 | Double | Density (no./m²) |
| Hier | | | | |
| | Organism | *Microtendipes pedellus* gr. | Text | Taxon name |
| | Phylum | Arthropoda | Text | Phylum |
| | Class | Insecta | Text | Class |
| | Order | Diptera | Text | Order |
| | Suborder | Nematocera | Text | Suborder |
| | Family | Chironomidae | Text | Family |
| | SubFamily | Chironominae | Text | Subfamily |
| | Tribe | Chironomini | Text | Tribe |
| | Genus | *Microtendipes* | Text | Genus |
| | Species | *Microtendipes pedellus* gr. | Text | Species |

**Table 18.** Example of the structure of a data file (Import_Format_Example_3.xls) in which the information needed to import the data is contained in two spreadsheets; one containing the sample information and a second containing the abundance and taxonomic hierarchy information.

[SampInfo contains the sample information. Abund_w_Hier contains the abundance and taxonomic hierarchy. The data in the two spreadsheets are linked by the Sample number and Sample columns. %, percentage; gr., group; no./m², number per square meter]

| Spreadsheet | Column name | Example data | Data type | Comment |
|---|---|---|---|---|
| SampInfo | | | | |
| | Sample number | ARBQR02 | Text | Sample identifier number |
| | Site number | 02341744 | Text | Site identifier number |
| | Stream name | Cat Creek | Text | Stream name |
| | Sampler | Hess | Text | Sampler |
| | Habitat | Riffle | Text | Habitat sampled |
| | Site abbrev. | ARBQR | Text | Site abbreviation |
| | Rep | 2 | Integer | Replicate sample number |
| | Date | 8/15/2002 | Date | Sample collection date |
| | Subsamp | 18.76 | Double | Subsample (%) |
| | Year | 2002 | Integer | Year of sample collection |
| Abund_w_Hier | | | | |
| | Sample | ARBQR02 | Text | Sample identifier number |
| | Taxon | *Microtendipes pedellus* gr. | Text | Taxon name |
| | Density | 10.661 | Double | Density (no./m²) |
| | Phylum | Arthropoda | Text | Phylum |
| | Class | Insecta | Text | Class |
| | Order | Diptera | Text | Order |
| | Suborder | Nematocera | Text | Suborder |
| | Family | Chironomidae | Text | Family |
| | SubFamily | Chironominae | Text | Subfamily |
| | Tribe | Chironomini | Text | Tribe |
| | Genus | *Microtendipes* | Text | Genus |
| | Species | *Microtendipes pedellus* gr. | Text | Species |

**Table 19.**    Example of the structure of a data file (Import_Format_Example_4.xls) in which the information needed to import the data is contained in three spreadsheets.

[SampInfo contains the sample information. Abund contains the abundance information. Hier contains the taxonomic hierarchy. The data in the SampleInfo and Abund spreadsheets are linked by the Sample Code and Sample columns. Data in the Abund and Hier spreadsheets are linked by the Taxon and Organism columns. %, percentage; gr., group; no./m², number per square meter]

| Spreadsheet | Column name | Example data | Data type | Comment |
|---|---|---|---|---|
| SampleInfo | | | | |
| | Sample Code | ARBQR02 | Text | Sample identifier number |
| | Site number | 02341744 | Text | Site identifier number |
| | Stream name | Cat Creek | Text | Stream name |
| | Sampler | Hess | Text | Sampler |
| | Habitat | Riffle | Text | Habitat sampled |
| | Site abbrev. | ARBQR | Text | Site abbreviation |
| | Rep | 2 | Integer | Replicate sample number |
| | Date | 8/15/2002 | Date | Sample collection date |
| | Subsamp | 18.76 | Double | Subsample (%) |
| | Year | 2002 | Integer | Year of sample collection |
| Abund | | | | |
| | Sample | ARBQR02 | Text | Sample identifier number |
| | Taxon | *Microtendipes pedellus* gr. | Text | Taxon name |
| | Density | 10.661 | Double | Density (no./m²) |
| Hier | | | | |
| | Organism | *Microtendipes pedellus* gr. | Text | Taxon name |
| | Phylum | Arthropoda | Text | Phylum |
| | Class | Insecta | Text | Class |
| | Order | Diptera | Text | Order |
| | Suborder | Nematocera | Text | Suborder |
| | Family | Chironomidae | Text | Family |
| | SubFamily | Chironominae | Text | Subfamily |
| | Tribe | Chironomini | Text | Tribe |
| | Genus | *Microtendipes* | Text | Genus |
| | Species | *Microtendipes pedellus* gr. | Text | Species |

Once the three data sources have been opened (abundance, sample information, and taxonomic hierarchy), the program looks for an Access database that contains information on data conversion formats. By default, the IDAS program expects to find this information in the file Data_Formats.mdb (or Data_Formats.accdb), which is placed in the same directory as the IDAS program (IDAS.exe) when IDAS is installed. If the program does not find this file in that location or if the file does not contain the required data tables, the program will give the user the option of searching another location for the data formats, creating a new data format file, or proceeding without predefined data formats (fig. 22). The user can use this feature to create multiple data format conversion files and store them in locations other than the program file directory (for example, in directories containing the data files) by removing the default conversion format file (Data_Formats.mdb) from the program directory or renaming the file.

When the data source and conversion format files have been successfully opened, the main screen of the format conversion procedure will be displayed (fig. 23). The File and Table information for the Abundance data, Sample/site information, and Taxonomic hierarchy are automatically populated with the file and spreadsheet/table names used to supply these data. The abundance (❸), sample information (❹), and taxonomic hierarchy frames (❺) are not displayed until the links between the source data tables (❷) have been successfully established or a data conversion format has been selected from the Formats menu (❻).

The links between the Abundance data and the Sample/site information and Taxonomic information (❷) are populated by selecting variables from the associated drop-down lists, which list all the variables (columns) in the spreadsheet or table. The fields that link tables must be of the same data type (for example, integer, string, date) though they may have different names (for example, **Sample** and **Sample number**, fig. 23). If the fields have different data types, IDAS will display a warning message and will reset the linked fields to blank. If the linked fields are of the same data type, IDAS will calculate how many of the rows in the abundance data table have corresponding values in the linked sample information or taxonomic hierarchy table. If there are entries in the abundance data table that do not correspond to the linked sample information or taxonomic hierarchy table, IDAS will display a warning message and will reset the linked fields to blank because a properly linked table will have corresponding values for all of the abundance data. While checking the data types and correspondence of fields in linked tables are valuable error-checking tools, it is still possible to make links that are not valid. The best way to avoid this is to use the same field names for fields that should be linked (for example, "Sample number" in the abundance data and sample information tables and "Taxon" in abundance data and taxonomic hierarchy tables). Ultimately, the user is responsible for correctly linking tables.

Once the links between the data sources have been successfully established, the abundance (❸), sample information (❹), and taxonomic hierarchy frames (❺) will be displayed and the drop-down lists will be populated with the column names from the associated source files. Initially, the sample information and taxonomic hierarchy frames are inactive (dimmed). The sample information frame will be activated when the abundance frame has been populated and the taxonomic hierarchy frame will be activated when the sample information frame has been populated. The fields in the drop-down lists are used to map the columns in the user-supplied source tables to the columns in the Bio-TDB formatted file. For example, selecting Taxon from the BU_ID drop-down list will map the contents of the Taxon column to the BU_ID column in the Bio-TDB formatted file. Only variables in the drop-down list can be selected, with two exceptions— SUID and Reach— which are distinguished by their shaded entry fields. The user has the option of typing data into these fields, for example, typing in a four-character SUID. In addition, the SampleID, SortCode, SUID, Sample Type, and SMCOD drop-down lists have additional options at the top of the lists that can be used to create variables from combinations of other variables. For example, the option Create SampleID can be used to create the numeric Bio-TDB SampleID from a combination of sample and site identifiers.

The fields in the abundance (❸), sample information (❹), and taxonomic hierarchy frames (❺) can be populated manually by selecting values from the drop-down lists or automatically by selecting a file conversion format from the Formats menu (Formats/Select a format) or by clicking on the Populate with default values

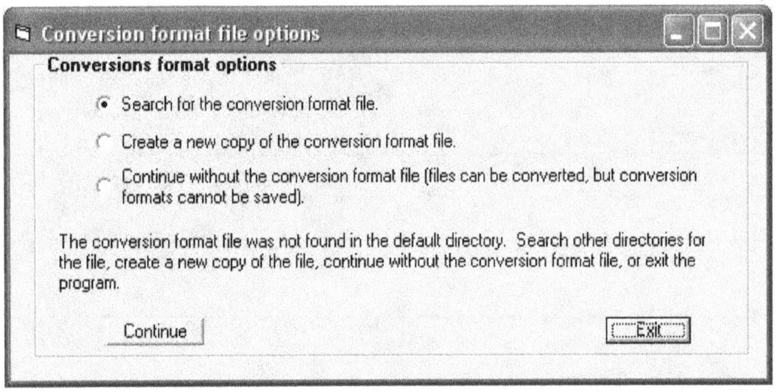

**Figure 22.**   The Conversion format file options window appears when the IDAS program cannot find the default conversion format file (Data_Formats. mdb) in the program directory.

[The user has the option of opening a conversion format file that exists elsewhere, creating a new copy of the conversion format file, or continuing without a conversion format file. The Search and Create a new copy procedures allow the user to name the conversion format file to something other than "Data_Formats.mdb" though the conversion format file must be an Access database]

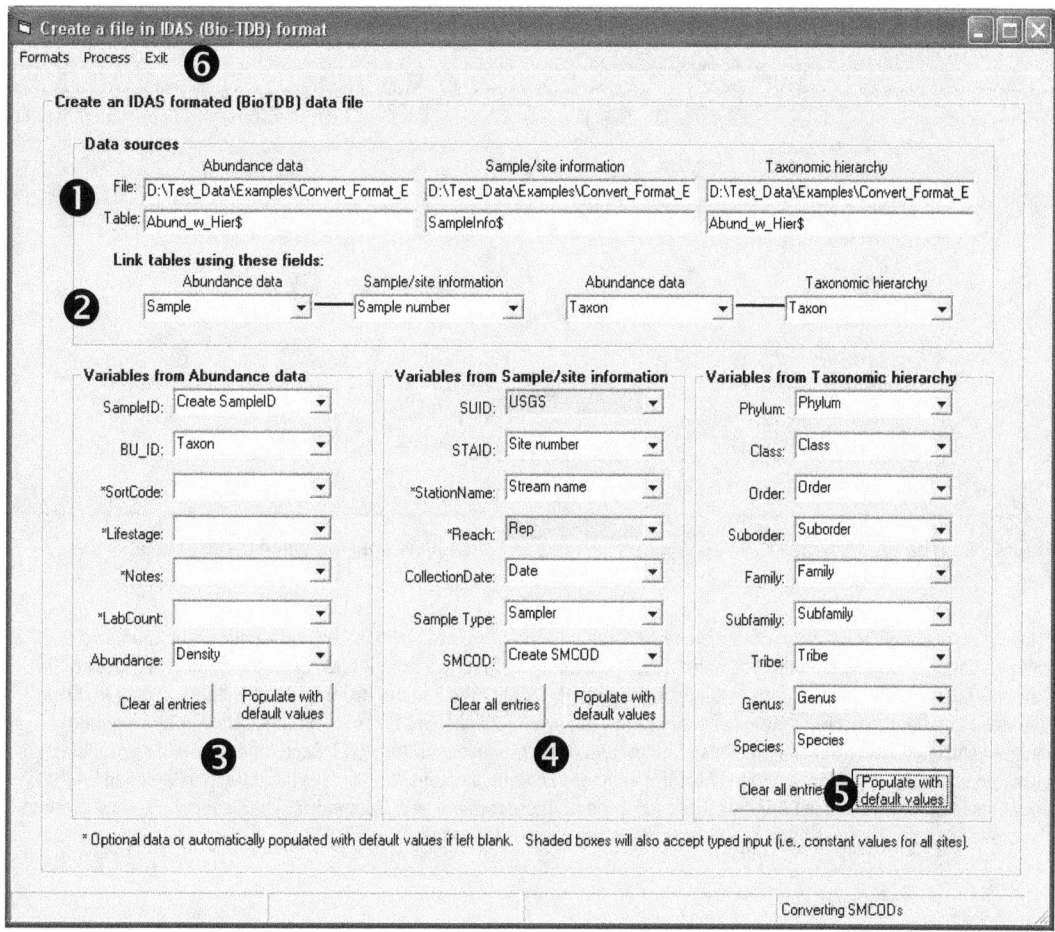

**Figure 23.**    Main screen of the format conversion procedure. Information entered here is used to convert data files to the format (BIO-TDB) used by IDAS.

[❶ List of files and data tables that provide information on abundance, sample information, and taxonomic hierarchy. ❷ Data fields that provide links between the source data tables. ❸ Drop-down lists of variables associated with abundances of each taxon. ❹ Drop-down lists of variables associated with sample and site information. ❺ Drop-down lists of taxonomic levels within the taxonomic hierarchy. ❻ Menus used to select previously saved conversion formats or create and save new conversion formats (Format), convert the data to IDAS format (Process), or exit the file conversion procedure (Exit)]

button. Information entered into the abundance (❸), sample information (❹), and taxonomic hierarchy (❺) fields can be cleared using the **Clear all entries** button or by double clicking on the label (for example, *SortCode) associated with each field (fig. 23). The **Clear all entries** button clears all field entries in the frame. Double clicking the label clears only the associated field. The IDAS program contains a variety of error-checking routines that check the values entered into each field. Error checks are conducted each time a value is manually selected from a drop-down list or when fields are automatically populated by selecting a file conversion format (Select a format). Clicking on the **Populate with default values** button populates the fields but error checks are not performed until the Process menu item is selected. The

definition, data requirements, and conversion process for each field are detailed in the following paragraphs.

**SampleID:** IDAS uses a numeric (positive integer) value to identify unique samples. If the source data contains such an identifier, it can be selected from the drop-down list. If samples are identified by another variable (for example, an alphanumeric sample identifier) or by a combination of variables (for example, station + date + sample type + replicate number), then the **Create SampleID** option should be selected. Selecting this option brings up the Create a unique sample identification number window (fig. 24), which allows the user to create SampleIDs starting with a specific number (positive integer, ❶) and one or more variables that uniquely identify each sample (❺) from a list of the available

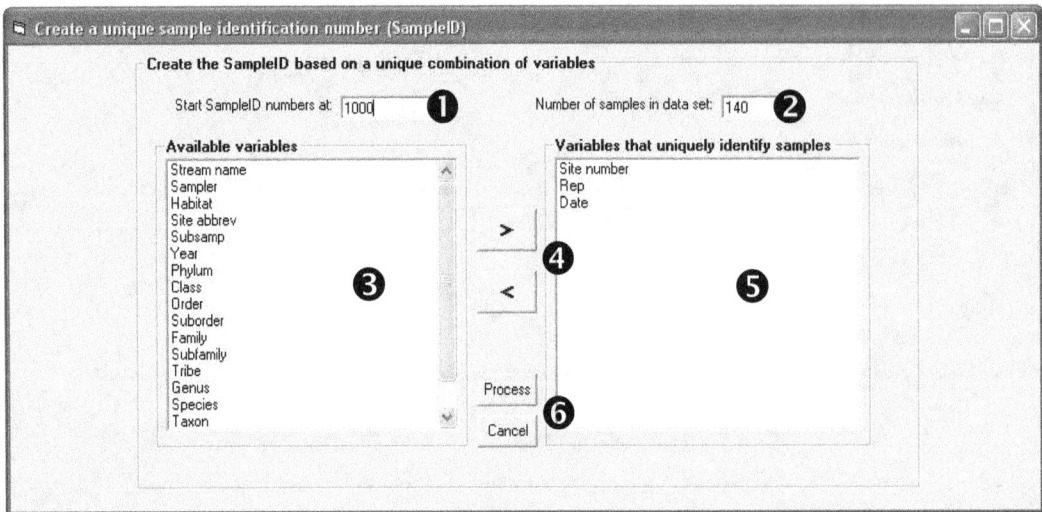

**Figure 24.** The Create a unique sample identification number window appears when the user selects the Create SampleID option in figure 23.

[This window allows the user to create a unique SampleID (positive integer) based on an existing alphanumeric sample identifier or a combination of identifiers. The user must specify a number (❶) at which to start numbering the SampleIDs and select one or more variables that uniquely identify the sample (❺) from the available variables (❸). Variables can be transferred between list boxes (❸, ❺) by double clicking an entry in the window or by selecting multiple entries (Ctrl- or Shift+right mouse button) and using the arrow buttons (❹). Each time a variable is transferred between list boxes, the program calculates the number of unique samples that exist in the dataset (❷) based on the variables in the right-hand list box (❺). The Process button creates the new SampleIDs. The Cancel button will cancel the creation of the SampleID and return the user to the convert formats screen (fig. 23)]

sample identifiers (❸). Variables can be transferred between the list boxes (❸,❺) by double clicking an entry in the window or by selecting multiple entries (Ctrl+right mouse button or Shift+right mouse button) and using the arrow buttons (❹). Every time a variable is transferred between list boxes, the program calculates how many unique samples exist in the dataset (❷) based on the variables in the right-hand list box (❺). Clicking on the Process button (❻) creates the new SampleIDs. Clicking on the Cancel button (❻) cancels the creation of the SampleID and returns the user to the Create a file in IDAS (Bio-TDB) format screen (fig. 23).

**BU_ID:** The BU_ID is the name given to the organism by the USGS NWQL biological unit. Select a variable from the drop-down list that corresponds to the name of the organism.

**SortCode (optional):** The SortCode is an integer value that is used to sort the sample data into an approximation of phylogenetic order. SortCodes can be populated by selecting a variable from the drop-down list that corresponds to the SortCode of the organism, selecting Create SortCode from the drop-down list, or leaving the field blank. The SortCodes created by IDAS arrange taxonomic information in the order presented in table 7 and is unique to each dataset.

**Lifestage (optional):** The lifestage field (text) is used to identify adult (A), larval (L), and pupal (P) life stages. If a variable holding lifestage data is selected from the drop-down list, the variable must identify life stages as "A," "L," or "P." If this field is left blank, the Lifestage column in the Bio-TDB formatted file will also be blank.

**Notes (optional):** The Notes field (text) is used to enter information on specimen condition that may be used to determine if taxa are retained in the dataset (Data Preparation module). IDAS recognizes the following entries for Notes (Moulton and others, 2000): imm., dam., indet., gender, artifact, or mount (multiple entries should be separated by commas). IDAS will alert the user if it does not find any notes that it can interpret. If this field is left blank, the Notes column in the Bio-TDB formatted file will also be blank.

**LabCount (optional):** The LabCount (double) was originally used to hold information on the counts generated by laboratory processing and used to calculate abundances. The IDAS program does not use the LabCount field in its calculations.

**Abundance:** The number of individuals of a taxon (BU_ID) that were found in the sample after correcting for laboratory and field subsampling. Abundance can be equivalent to density if the data have been corrected for sampler size and number. There can be multiple lines of abundance for a BU_ID in a sample if the BU_ID contains multiple lifestage and (or) multiple laboratory processing notes (Notes). The abundance for organisms in a qualitative sample should be set to 1.

**SUID:** The Study-Unit identifier is a 4-character project identifier that is used in the formation of the sample code (SMCOD). The SUID can be supplied by a column in the Sample Information data, extracted from the SMCOD (if the data contains a valid SMCOD), or entered as a 4-character constant (for example, USGS) that will be applied to all the data in the dataset. The SUID "EMAP" is reserved for data imported in WR-EMAP format and should not be used for data that are converted to Bio-TDB format.

**STAID:** The station identifier is a text string that uniquely identifies the site (location on a body of water) where samples are collected. The STAID in NAWQA datasets is the official USGS station number. In non-NAWQA datasets, this could be a short name or number that identifies the body of water (also see StationName).

**StationName (optional):** A text string containing the name of the station from which the sample was collected. Typically, this is a descriptor of the water body and nearby geographic references, for example, "Stony Brook at Hwy 140 nr St. Louis, VT."

**Reach (optional):** A text string used to designate the location within a site from which a sample was collected. In the NAWQA Program, the Reach corresponds to a length of stream from which samples are collected and composited. In non-NAWQA datasets, this field can be used to represent a sampling location within a site or one of a series of replicate samples. The Reach field can be populated from a field in the Sample Information dataset or by entering a constant value that will be used to identify Reaches in all rows of data. If the Reach field is left blank, IDAS will automatically identify all data rows as Reach "A."

**CollectionDate:** A date/time field that is used to identify the calendar date on which the sample was collected. This field must conform to a valid date format used by Excel or Access.

**SampleType:** A text field that is used to identify samples that are qualitative (presence/absence) or quantitative. Sample types can be supplied by a field in the sample information dataset, extracted from the SMCOD (if the sample information dataset contains a valid SMCOD), or set to a constant

**TIP: The use of "EMAP" as a SUID is restricted to data imported as WR-EMAP data. "EMAP" should not be used as a SUID for data imported into IDAS by other means.**

value ("All qualitative" or "All quantitative") if the dataset is composed of only one data type. The IDAS program can only differentiate between two sample types, "Q" for qualitative and "R" for quantitative. Consequently, the Extract from SMCOD option can only recognize "Q" and "R" sample types as the sample type identifier in the SMCOD. If a Sample Information field is used to supply sample type information, the Identify quantitative and qualitative sample types window (fig. 25) is displayed. The arrow keys (❷,❹) are used to transfer the sample types in the Available Sample Types list box (❸) into the Quantitative sample types (❶) or Qualitative sample types (❺) list boxes. The Process and Cancel buttons (❻) are used to extract the sample types or to cancel the creation of sample types and return to the format conversion window (fig. 23).

**SMCOD:** The sample code is a text field that was used to identify samples in the field before assignment of the unique SampleID. This code consists of the 4-character SUID, month (MM) and year (YY) the sample was collected, "I" for invertebrate samples, "R" or "Q" to indicate the sample type (quantitative and qualitative, respectively), "M" to indicate that this is the main portion of the sample, and a 4-digit sequential number. The SMCOD can be supplied from a field in the Sample Information data or it can be generated (Create SMCOD) from the SUID, CollectionDate, and Sample Type fields. Because creation of the SMCOD is dependent on these fields, the SMCOD field will not be enabled until information is entered for these fields.

**Phylum-Species:** Text fields used to identify the elements of the taxonomic hierarchy associated with the organism for which abundance is reported (BU_ID). The taxonomic hierarchy does not have to supply all the taxonomic fields indicated in ❺ (fig. 23). However, the number of taxonomic hierarchy fields that are populated will affect the ability to set taxonomic levels and resolve taxonomic ambiguities in the Data Preparation module.

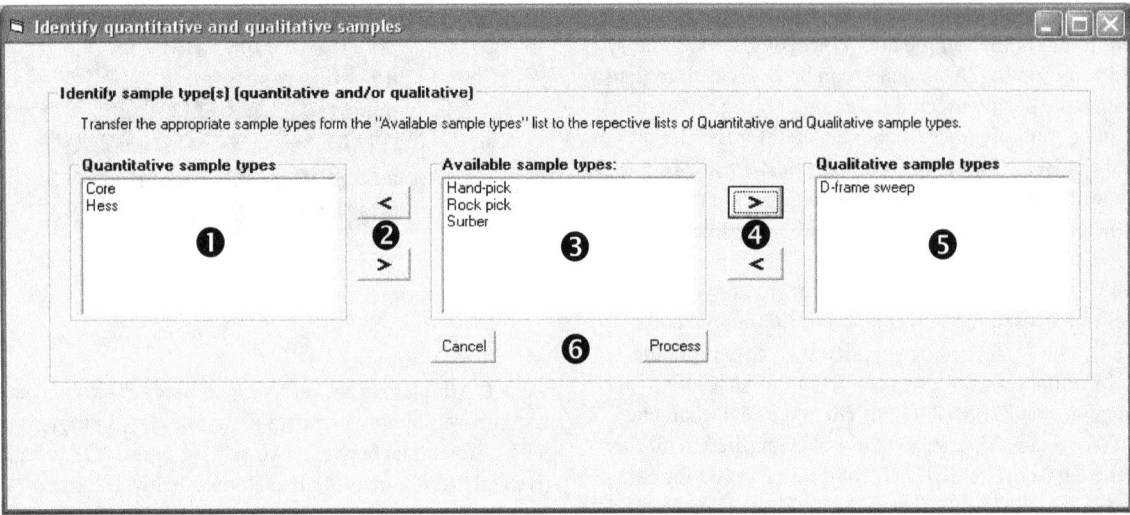

**Figure 25.**    The Identify quantitative and qualitative samples window allows the user to identify which sample types are quantitative and which are qualitative.

[The arrow buttons (❷,❹) are used to move sample types from the Available sample types list box (❸) to either the Quantitative sample types (❶) or Qualitative sample types (❺) list boxes. The arrow buttons support single or multiple selections. The Process button (❻) assigns the qualitative or quantitative identifier to the IDAS formatted file. The Cancel button (❻) cancels the assignment of sample types and returns the user to the conversion formats screen (fig. 23)]

## Using User-Defined Formats

Information on the table links and mapping of variables can be stored in an Access database and recalled for later use. The default name for the file of conversion formats is Data_Formats mdb and is stored in the IDAS program directory when IDAS is installed. The user can create new conversion format files with a different name, but the Data_Formats mdb file must first be removed from the IDAS program directory or renamed. The version of Data_Formats mdb provided with the IDAS program contains data conversion formats that will work for the four user-defined examples (tables 16–19). The conversion format database provided with IDAS contains four access format tables: (1) tblFormat contains the table links and variable mappings, (2) tblQual contains the list of qualitative sample identifiers, (3) tblQuant contains the list of quantitative sample identifiers, and (4) tblSampleID contains information on the number of identifiers used to create the SampleID (first row), the field chosen to represent the SampleID, and the list of field names that are combined to form the SampleID if the Create SampleID option was selected. These format tables will be populated when the conversion of the user-defined tables is executed.

**Selecting a user-defined format:** Existing formats can be selected from the Formats menu once the data tables and conversion format database have been opened (fig. 23). The Select a format submenu calls up the conversion format selection frame (fig. 26). Once an existing format has been selected from the drop-down list, the table links and variable mappings

will be entered into the appropriate fields and checked for validity. If the program encounters any invalid entries, the user will be notified and the entry will be cleared (valid entries will remain). The Create a unique sample identification number window (fig. 24) is automatically populated from the conversion format database when the Create SampleID option is part of the conversion format. The user needs to enter a starting number for the SampleID that will produce unique SampleIDs and that will not conflict with other datasets. The Identify quantitative and qualitative samples window (fig. 25) will also appear if multiple sample type identifiers are part of the data format conversion. The user needs to review this information, make any necessary changes, and proceed with processing.

**Figure 26.**    The Select a conversion format frame allows the user to select a predefined conversion format from the drop-down list.

**Defining a new format:** New formats can be defined by selecting the Define new format submenu from the Formats menu (fig. 23). This will clear any existing links or variable mapping information. Selecting the Define new format option sets a flag in the program that indicates that a new format exists and that the user should not be allowed to exit the program without being given the option of saving the new format.

**Saving a new or modified format:** The Save current format submenu can be used to save conversion information (linking fields and variable mappings) that can be used to process additional data files. This submenu of the Formats menu (fig. 23) is enabled once a dataset has been successfully converted. The Save current format submenu calls up a small window (fig. 27) with a drop-down list that contains the currently available file conversion formats. Entering a new name will create a new format entry. Entering an existing name will replace the old conversion format information with the new information. If the conversion format has changed or is new and the user attempts to exit the program without saving the conversion format, the program will issue a warning and will provide an opportunity to save the new or modified conversion format.

## Executing Conversions

Once the fields in figure 23 have been populated, selecting Process from the main menu will convert the dataset into a Bio-TDB formatted file. The program does a considerable amount of error checking during the conversion process and will inform the user if it encounters any problems in the data conversion. For some data conversion problems, IDAS will offer the user the option of allowing IDAS to resolve the problem or exiting the program and manually fixing the problem. For example, if the program detects multiple taxonomic

**Figure 27.** The Save conversion format as frame allows the user to enter a new name or select an existing name under which to save a conversion format.

hierarchies associated with a taxon, it can automatically select the most frequently occurring hierarchy and continue with the conversion process. Similarly, if multiple taxa are associated with a SortCode or multiple SortCodes are associated with a taxon, IDAS can generate new sort codes to resolve this problem.

If the user selected the Create SampleID option, the data in the resulting Bio-TDB formatted file will be identified by the new SampleID and not the identifier or combination of identifiers that were present in the source data file. Consequently, there will be no easy way to link the new SampleID to the original data. To maintain a link between the original and processed datasets, the IDAS program adds a field called "SampleID" to the source data table or spreadsheet that supplied the sample information. If the program detects that a field called "SampleID" already exists in the table or spreadsheet, it will inform the user and request a new field name under which to store the new SampleID (fig. 28).

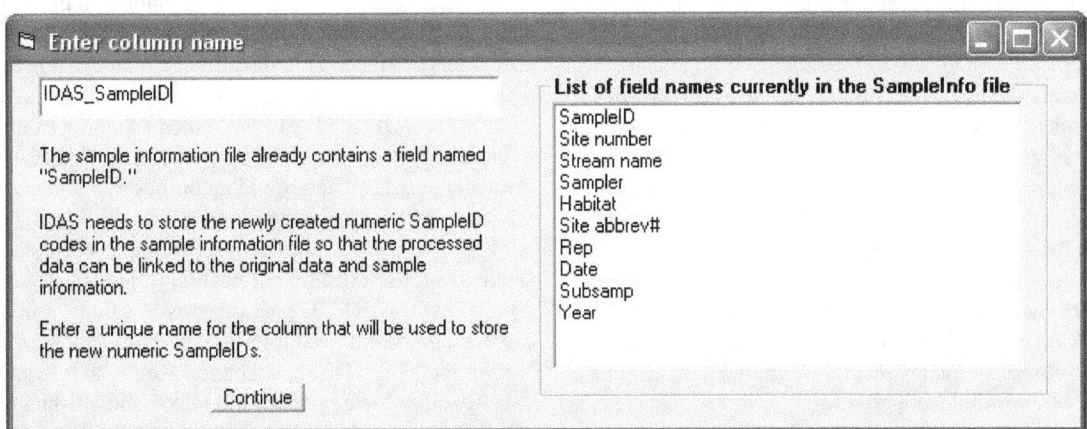

**Figure 28.** The IDAS program maintains the link between the original and converted datasets by adding the SampleIDs that it creates to the table or spreadsheet that was the source of the sample information.

[If the field "SampleID" already exists, this window appears and the user is asked to enter a new field name]

## Maintain Attributes File

The attribute file supplied with IDAS (Attributes_BEHAV_v5a.xls) contains five spreadsheets (table 20). The EQTX spreadsheet stores information on equivalent taxa for tolerance (EQTXTOL) and functional groups (EQTXFG) along with the BU_ID, NAME (BU_ID without authority name), Lifestage, SortCode, and date when the EQTX values were modified (MODIFIED). The ATTRIB spreadsheet stores the attribute information—five regional tolerance values, a national tolerance value, functional groups, and behavioral traits (table 21) compiled from Barbour and others (1999) and North Carolina Department of Environment and Natural Resources (2006). The HIER spreadsheet contains the taxonomic hierarchy for the taxa in EQTX. The Version spreadsheet stores information on how the attribute file was created. The Notes spreadsheet contains information on the derivation of the original attributes file. The IDAS program provides tools (table 22) for maintaining and creating versions of the attribute file that are tailored for particular datasets and regions of the country.

The attribute file provides tolerance data for five regions of the country (Mid-Atlantic, MATL_TOL; Southeast, SE_TOL; Upper Midwest, UMW_TOL; Midwest, MW_TOL; and Northwest, NW_TOL) along with a national tolerance value (NAT_TOL) that is the mean of the regional tolerance values. When calculating tolerance and functional group metrics, the IDAS program can match taxa to tolerance values and functional groups based on either the taxon name (NAME field) or the equivalent taxon (EQTXTOL for tolerance, EQTXFG for functional groups). Equivalent taxa are created by advancing up the taxonomic hierarchy from species to phylum to find tolerance and functional group data. For example, if there are no tolerance data reported for *Baetis tricaudatus*, the program would check for tolerance data at each level of the taxonomic hierarchy (genus, tribe, subfamily, family, order) until it finds a tolerance value or reaches the top of the taxonomic hierarchy (phylum). The first taxon with tolerance data (for example, *Baetis* sp.) would become the EQTXTOL name for *Baetis tricaudatus*. If no tolerance data are found, the EQTXTOL would be blank for *Baetis tricaudautus*. The EQTXTOL and EQTXFG taxa depend on the taxa that are in the dataset and the region of the country that is supplying the tolerance values. The attribute file that is supplied with IDAS (Attributes_BEHAV_v5a.xls) is not optimized for any particular region of the country or for a particular dataset. Therefore, if the user plans on calculating tolerance or functional group metrics using EQTXTOL and (or) EQTXFG, the attribute file must be optimized for the taxa in the dataset and for the region that is supplying the tolerance values. The tools provided in Maintain attributes file provide quick and efficient mechanisms for optimizing the attribute file.

The options for creating a new attribute file (Create a new attribute file) or modifying an existing attribute file (Modify an existing file) are located within the Maintain attributes submenu of the Files menu (figs. 29–30). The five options for creating and modifying attribute files have different requirements for source files (abundance data files, attribute files, tolerance files) and produce different outputs (table 22).

Two options are available for creating a new attribute file (fig. 29; table 22). New attribute files can be created on the basis of the taxa in an abundance file (Extract taxa from an abundance file) or the taxa in the taxonomic hierarchy table (HIER) of an existing attribute file (Extract taxa from HIER in attribute file). Extracting taxa from an abundance spreadsheet or table produces an attribute file that only contains information for the taxa that are present in the spreadsheet or table. Extracting taxa from the HIER spreadsheet produces an attribute file that contains all of the taxa that are present in the HIER spreadsheet. Extracting taxa from an abundance file provides a mechanism for producing an attribute file that is optimized for a particular region of the country and for a particular dataset. Extracting taxa from a HIER spreadsheet provides a mechanism for updating the EQTXTOL of the attribute file for a particular region of the country but not for a particular dataset. Both methods update the Version spreadsheet of the attribute file.

Three options are available for modifying an existing attribute file (fig. 30; table 22). The Add taxa from an abundance file option modifies an existing attribute file by adding taxa from an abundance file. New taxa are added to the HIER and EQTX spreadsheets, the equivalent taxa fields (SortCode, BU_ID, Lifestage, Name, EQTXTOL, EQTXFG, and Modified) are updated to include the new taxa, and the Version spreadsheet is updated with the information on the source data and date of the update. The Refresh an EQTX spreadsheet option updates the EQTX spreadsheet in an existing attribute file. This option is most often used to update the equivalent taxa (EQTXTOL and EQTXFG) when changing from one tolerance source (MATL_TOL) to another (SE_TOL) or after manually or automatically adding or modifying taxa in the ATTRIB spreadsheet of the attributes file. The Update the ATTRIB spreadsheet option is used to update (replace or supplement) the attribute information (tolerance or functional groups) in the ATTRIB spreadsheet with values from an external source. For example, the tolerance data for the Northwest (NW_TOL) could be replaced with tolerance data obtained from State agencies in Oregon and Washington. It is important to note that the Update the ATTRIB spreadsheet option does not automatically update the EQTX spreadsheet. Any of the other Maintain attribute file options can be used to update the EQTX spreadsheet by creating a new attribute file or modifying the existing attribute file.

> **TIP: The attribute file distributed with IDAS is generic and should not be used without optimizing the EQTX spreadsheet for the dataset and region.**

**Table 20.** The attributes workbook is composed of five spreadsheets.

[EQTX holds equivalent taxa information. ATTRIB holds the functional group, tolerance, and behavioral traits information. HIER holds the taxonomic hierarchy for each taxon. Version holds information on when and how the attribute file was created. Notes holds information on how the original attributes file was populated. -999, missing value]

| Table | Column | Data type | Example | Comments |
|---|---|---|---|---|
| EQTX | | | | Equivalent taxa spreadsheet |
| | SortCode | Long | 974 | Sort code |
| | BU_ID | Text | *Perlesta shubuta* Stark | Taxon name |
| | Lifestage | Text | L | Lifestage (L, P, A, or blank) |
| | NAME | Text | *Perlesta shubuta* | Taxon without authority name |
| | EQTXTOL | Text | *Perlesta* | Equivalent taxa for tolerance |
| | EQTXFG | Text | *Perlesta* | Equivalent taxa for functional group |
| | MODIFIED | Date/time | 1/29/2004 | Date created or modified |
| ATTRIB | | | | Attributes spreadsheet |
| | SEQ | Long | 1589 | Sequence number for sorting |
| | NAME | Text | *Perlesta* | Taxon |
| | NAT_TOL | Double | 4.7 | National tolerance |
| | SE_TOL | Double | 4.7 | Southeast tolerance |
| | UMW_TOL | Double | -999 | Upper mid-west tolerance |
| | MW_TOL | Double | 4.5 | Mid-west tolerance |
| | NW_TOL | Double | 4.5 | Northwest tolerance |
| | MATL_TOL | Double | -999 | Mid Atlantic tolerance |
| | FG | Text | PR | Functional group code |
| | BEHAV | Text | cn | Behavior trait code |
| HIER | | | | Taxonomic hierarchy spreadsheet |
| | SortCode | Long | 974 | Sort code |
| | BU_ID | Text | *Perlesta shubuta* Stark | Taxon name with authority name |
| | TxLvl | Integer | 9 | Taxonomic level: nine = species |
| | Name | Text | *Perlesta shubuta* | Taxon name without authority name |
| | Phylum | Text | Arthropoda | Phylum |
| | Class | Text | Insecta | Class |
| | Order | Text | Plecoptera | Order |
| | Suborder | Text | Systellognatha | Suborder |
| | Family | Text | Perlidae | Family |
| | Subfamily | Text | Acroneuriinae | Subfamily |
| | Tribe | Text | Acroneuriini | Tribe |
| | Genus | Text | *Perlesta* | Genus |
| | Species | Text | *Perlesta shubuta* | Species |
| | Modified | Date/time | 1/28/2004 | Date created or modified |
| Version | | | | Creation information |
| | Created | Text | 7/24/2008 | Date created or modified |
| | Attribute_source_file | Text | Attrib_SE.xls, ATTRIB$ | Source of attribute file |
| | Tolerance_source | Text | NW_TOL | Regional tolerance source |
| | Abund_source_file | Text | D:\IDAS_DPAC.xls | File providing taxa |
| | Abund_table | Text | DPAC | Table or spreadsheet providing taxa |
| Notes | | | | Notes on attributes file |
| | No | Text | 1 | Sequential number for note |
| | Date | Date/time | 1/14/2002 | Date created |
| | Comments | Text | *Falloporus* to *Lioporeus* | *Falloporus* changed to *Lioporeus* |

**Table 21.**    Functional group and behavioral traits are categorized in the ATTRIB spreadsheet.

| Trait | Abbreviation | Explanation |
|---|---|---|
| Functional groups | | |
| | PA | Parasites |
| | PR | Predators |
| | OM | Omnivores |
| | GC | Collector-gatherers |
| | FC | Filtering-collectors |
| | SC | Scrapers |
| | SH | Shredders |
| | PI | Piercers |
| Behavioral | | |
| | cn | Clinger |
| | cb | Climber |
| | sp | Sprawler |
| | bu | Burrower |
| | sw | Swimmer |
| | dv | Diver |
| | sk | Skater |

**Table 22.**    The Maintain attributes file submenu provides five options for creating new attribute files or modifying existing attribute files.

| Submenu options | Source files: Attribute file and... | | Allow changes in tolerance source? | Output file |
|---|---|---|---|---|
| | Abundance | TOL or FG* | | |
| Create a new attribute file | | | | |
| 1. Extract taxa from an abundance file | Yes | No | Yes | New attribute file |
| 2. Extract taxa from HIER in attribute file | No | No | Yes | New attribute file |
| Modify an existing attribute file | | | | |
| 3. Add taxa from an abundance file | Yes | No | No | Original attribute file with new taxa added to EQTX and HIER |
| 4. Refresh an EQTX spreadsheet | No | No | Yes | Original attribute file with revised fields in EQTX spreadsheet |
| 5. Update the ATTRIB spreadsheet | No | Yes | Yes | Original attribute file with tolerance or functional group data added and (or) replaced from an external file |

* Incorporate tolerance or function group data from an external data file.

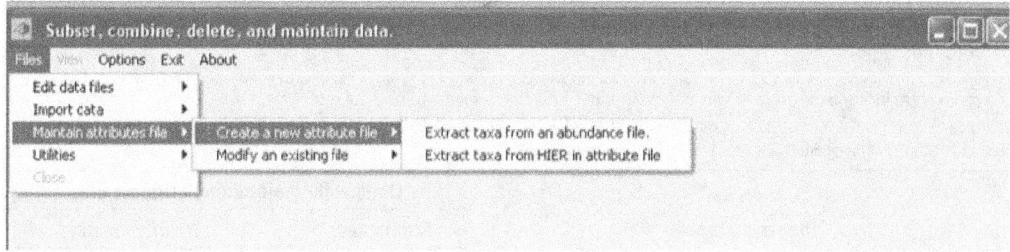

**Figure 29.** A new attribute file can be extracted taxa from an abundance file or the HIER spreadsheet of an attribute file.

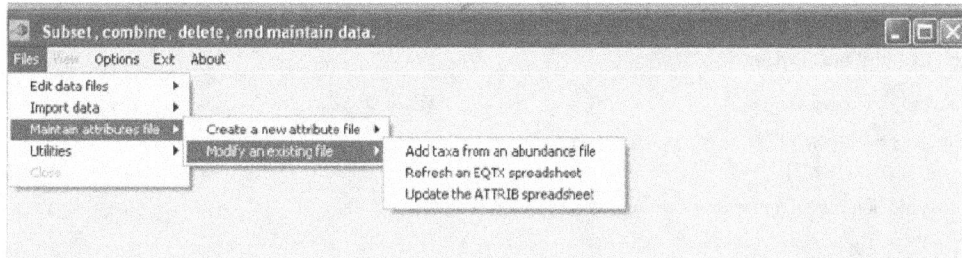

**Figure 30.** An existing attribute file can be modified by adding taxa from an abundance file, refreshing an EQTX spreadsheet, or updating the ATTRIB spreadsheet.

## Creating or Modifying Attribute Files

The process of creating or modifying attribute files begins by selecting one of the five processing options listed under the Maintain attributes file submenu (figs. 29, 30). The sequence of screens that are displayed while processing the attribute file depends on the option selected (table 23). Most of these screens are standard screens used to open or save a file (fig. 5) or spreadsheet or data table (fig. 6). The attribute maintenance functions that derive data from abundance files (options 1 and 3) can use data from both Bio-TDB (table 1) and processed (table 3) data files. Consequently, the user must specify the type of data to process (fig. 12) for options 1 and 3. All options except option 3 (add taxa from an abundance file to an existing attribute file) require that the user specifies the source of the tolerance values (five regions or national) that will be used to optimized the EQTXTOL values (options 1, 2, and 4) or identify the attribute (regional tolerance value or functional groups) of the ATTRIB spreadsheet that will be updated from an external source (option 5). The Options for Tolerance and Functional Group Metrics screen (fig. 31) is used to enter this information. When this screen is displayed under option 5, a radio button for functional groups (FG) is added allowing the user to add functional group data to the ATTRIB spreadsheet.

Options 1–4 automatically update the EQTX spreadsheet by looking for a match between the taxon and the NAME field of the ATTRIB spreadsheet. Because the NAME field of the ATTRIB spreadsheet does not include a taxonomic authority, the IDAS program removes authority names from the BU_ID before populating the NAME field of the EQTX spreadsheet. Taxa in the EQTX spreadsheet are identified by both the BU_ID and NAME fields. The IDAS program uses a three-part procedure to match taxa to attributes and to populate the EQTXTOL and EQTXFG columns. EQTXTOL values are derived using the following procedure:

1. Check for a match between the NAME fields in the EQTX and ATTRIB spreadsheets. If a match exists (that is, the NAME fields match and a tolerance value exists), then the EQTXTOL field is set to NAME. If a match cannot be found, the next procedure is used.

2. Check for a match between the BU_ID and the NAME fields in EQTX. If a match exists and a tolerance value exists for the NAME field in the ATTRIB spreadsheet, then set the EQTXTOL field associated with BU_ID to NAME. This procedure builds on previously established equivalencies. If the user has elected to create a new attribute file, then the EQTX spreadsheet is empty and this procedure will be skipped. If no match is found, the next procedure is used.

**Table 23.**    Processing screens that are displayed when using the five options for creating or modifying attribute files.

[The screens are listed in the order in which they appear. "Yes," the screen will be displayed. "No," the screen will not be displayed. The captions of the screens used to select files (fig. 5) and spreadsheets or tables (fig. 6) will change depending on whether files are being opened or saved]

| | Options for maintaining attribute files | | | | |
|---|---|---|---|---|---|
| | Create new | | Modify existing | | |
| **Processing screen** | **1** | **2** | **3** | **4** | **5** |
| Open abundance file (fig. 5) | Yes | No | Yes | No | No |
| Select raw or processed data (fig. 12) | Yes | No | Yes | No | No |
| Select table from abundance file (fig. 6) | Yes | No | Yes | No | No |
| Select tolerance region (fig. 31) | Yes | Yes | No | Yes | Yes* |
| Select source attribute file (fig. 5) | Yes | Yes | Yes | Yes | Yes |
| Enter new attribute file name (fig. 5) | Yes | Yes | No | No | No |
| Review EQTX entries (fig. 32) | Yes | Yes | Yes | Yes | No |
| Select external file holding tolerance or functional group data (fig. 5) | No | No | No | No | Yes |
| Select table holding tolerance or functional group data (fig. 6) | No | No | No | No | Yes |
| Select options for updating attributes (fig. 31) | No | No | No | No | Yes |

* A radio button for functional groups (FG) is added to the select regional tolerance screen (fig. 31) when modifying an attributes file using option 5.

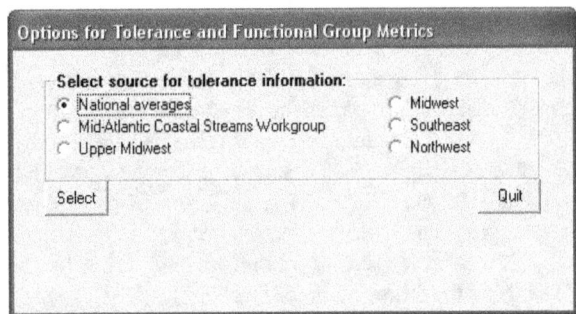

**Figure 31.**    The user can optimize the EQTXTOL data based on tolerance data from six sources.

3. Check the taxonomic hierarchy associated with the NAME field in the EQTX spreadsheet and see if any higher taxonomic levels produce a match with the NAME field in the ATTRIB spreadsheet. If a match is found (that is, the name associated with a taxa level matches the NAME field in the ATTRIB spreadsheet and a tolerance value exists), then set the EQTXTOL value to the taxon name at that level in the hierarchy. If no match can be found, set the EQTXTOL field to blank.

This procedure is also used to populate the EQTXFG field, which stores taxa that match functional group data in the ATTRIB spreadsheet. The procedures used to populate EQTXTOL and EQTXFG are independent and can result in entries for EQTXTOL and EQTXFG that are different from one another and different from the NAME field.

The procedures used to remove authority names in IDAS generally are accurate and reliable. However, the variability in how authority names are embedded within BU_IDs, particularly for provisional and conditional identifications, can lead to errors that obscure matches with the NAME field. Taxonomic revisions may also lead to problems in matching taxa with their attributes if the NAME fields in the ATTRIB spreadsheet do not include these revisions. To address these issues, IDAS includes a review function that allows the user to examine and edit the matches between taxa and their attributes (fig. 32).

## Review NAME Fields

After IDAS has made its assessment of the "best" matches between BU_IDs and attributes, the Review the NAME field form (fig. 32) will be displayed. This form allows the user to review and modify entries in the NAME field of the EQTX spreadsheet of the attributes file. The following information is displayed on this form:

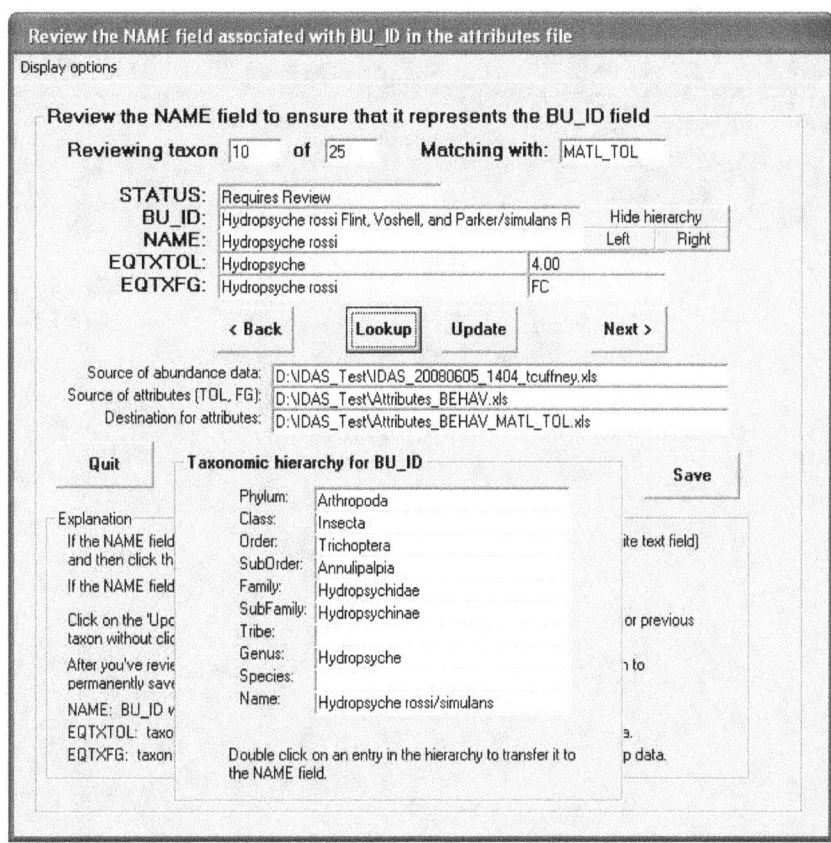

**Figure 32.** The Review the Name field form is used to review and modify the NAME field.

[This form is used to find matches with tolerance (EQTXTOL) and functional group (EQTXFG) data contained in the attributes file]

**Display options menu**: These options are used to limit the number of taxa that the user needs to review (fig. 33). A total of eight options are provided grouped under two main options:

1. All taxa: displays all of the taxa in the dataset.

2. Only taxa not in EQTX: displays only those taxa not currently listed in the EQTX spreadsheet of the source attribute file.

The following sub options are available under the main display options:

A. All STATUS values: displays all taxa regardless of the status value.

B. STATUS = "Requires review": displays only those taxa where the status value is set to "Requires review." This focuses the review on provisional/conditional identifications.

C. EQTXTOL or EQTXFG = "No match": displays only those taxa where IDAS could not find a match either for tolerance (EQTXTOL) or functional group (EQTXFG).

D. "No match" and "Requires review": displays only provisional/conditional taxa (Status = requires review) where IDAS could not find a match for either tolerance of functional group (EQTXTOL or EQTXFG = "No match").

**Reviewing taxon**: a counter that shows the number of the taxon that is being displayed and the total number of taxa being reviewed.

**Matching with**: an abbreviation that identifies the source (region) of the tolerance values (fig. 31).

**STATUS**: indicates the status of the data being displayed. **NEW** indicates that the information being displayed has not been previously viewed or modified. **Requires review** indicates that the BU_ID contains characters that suggest that the NAME automatically generated by IDAS may be derived from provisional/conditional identification and should be reviewed for accuracy. **Modified** indicates that the user has previously viewed and modified the NAME field. **Reviewed** indicates that the user has already viewed the data being displayed, but

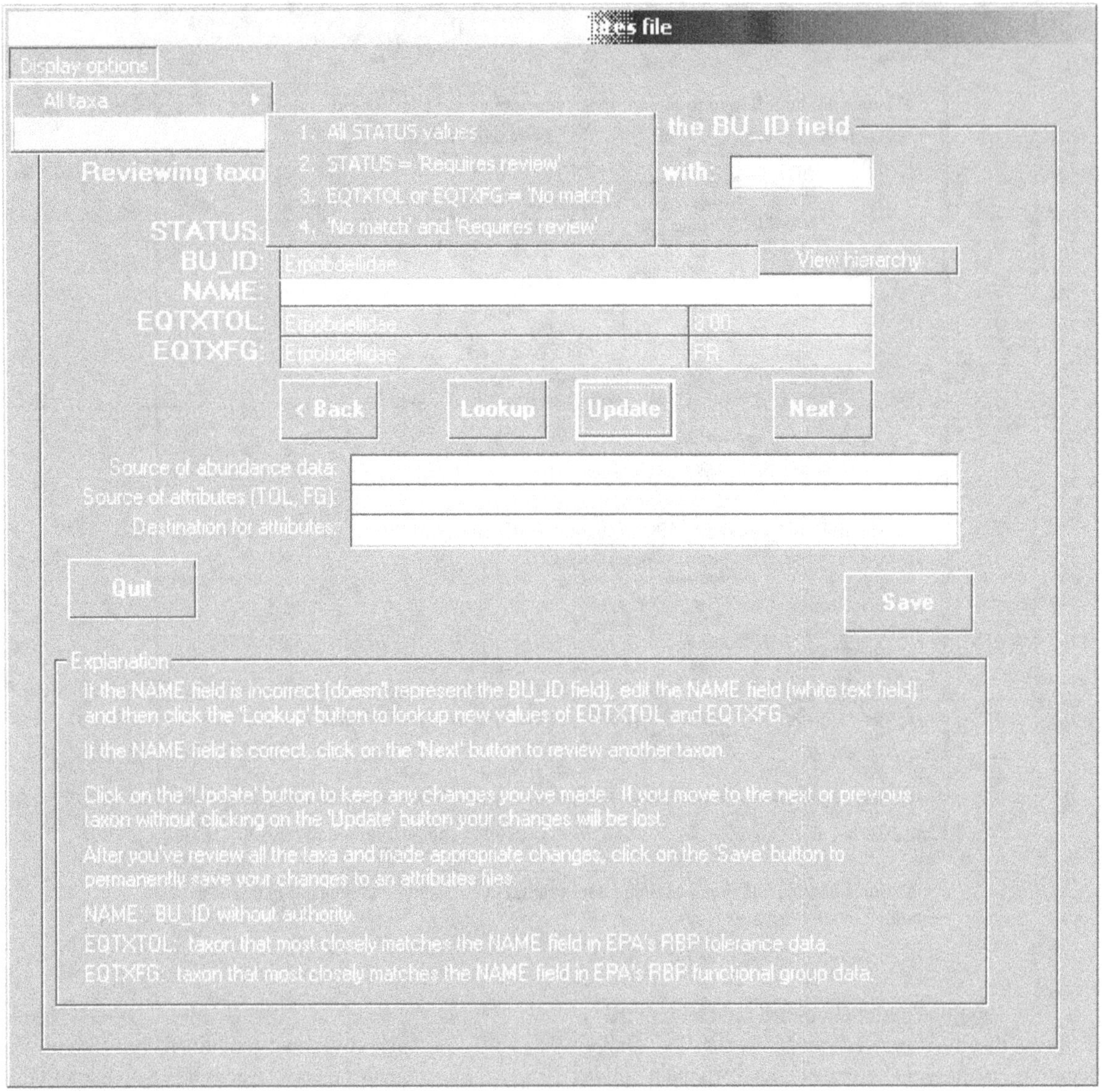

**Figure 33.**    Eight display options can be used to limit the number of taxa to review.

[Four sub options are available under the options All taxa and Only taxa not in EQTX]

did not modify it. The original status of the taxa is given in parentheses.

BU_ID: shows the name of the taxon along with the authority, if applicable. This field is highlighted in yellow indicating that this field cannot be edited.

View hierarchy or Hide hierarchy: displays or hides the taxonomic hierarchy associated with the BU_ID (see Taxonomic hierarchy for BU_ID; fig. 32).

NAME: corresponds to the BU_ID name after IDAS has removed the authority name. The NAME field is used to find matches with the tolerance, functional group, and behavioral

attributes in the ATTRIB spreadsheet. The primary purpose of the Review the NAME field form is to provide the user the opportunity to review and correct the NAME field. The NAME field, which is highlighted in white, is the only field on this form that can be edited.

Right/Left: provisional/conditional editing suggestions. Buttons displayed to the right of the NAME data line can be used to convert provisional/conditional identification to conventional identifications. In the example given in figure 32, the Right and Left buttons will extract the left and right taxa names (*Hydropsyche rossi* and *Hydropsyche simulans*, respectively). Other options for resolving provisional/conditional names will be displayed as appropriate. These buttons minimize the amount of typing required to resolve provisional/conditional names.

EQTXTOL: displays the name of the taxon that provided the closest match with the tolerance data contained in the spreadsheet ATTRIB. The number to the right of the EQTXTOL text box is the tolerance value associated with EQTXTOL. If IDAS could not find a match that produced tolerance data, then the EQTXTOL name and value will be listed as **No match**.

EQTXFG: displays the name of the taxon that provided the closest match with the functional group data contained in the spreadsheet ATTRIB. The abbreviation to the right of the EQTXFG text box is the functional group identifier associated with EQTXFG (table 21). If IDAS could not find a match that produced functional group data, then the EQTXFG name and abbreviation will be listed as **No match**.

Back: button that moves backward through the list of new taxa. Clicking on this button will bring up the previously viewed taxon and will change the STATUS field from **NEW** or **Requires review** to **Reviewed**. It will not change the STATUS field if it is already set to **Modified**.

Lookup: button that causes IDAS to find the "best" matches for tolerance and functional group information associated with the NAME field. Clicking on this button will update information in the EQTXTOL and EQTXFG fields (both names and values) using the three-part procedure described above.

Update: button that saves changes to the entries in the STATUS, BU_ID, NAME, EQTXTOL, and EQTXFG fields. If the Back or Next buttons are pressed without first pressing the Update button, no changes will be saved and the BU_ID, NAME, EQTXTOL, and EQTXFG fields will revert back to their previous values. The STATUS field will change from **NEW** to **Reviewed** to indicate that the user has reviewed this taxon without making changes. The Update button saves changes to a temporary recordset. It does not save changes to the attribute file.

Next: button that moves forward through the list of new taxa. Clicking on this button will change the STATUS field from **NEW** or **Requires review** to **Reviewed** and will then display the next taxon. It will not change the STATUS field if it is already set to **Modified**.

Source of abundance data: name of the file that provided the invertebrate abundance data.

Source of attributes: name of the file that provided the attributes data.

Destination for attributes: name of the file that will store the attributes data. If the user has selected the option to add

tolerances to an existing attribute file, the source and destination file names will be the same.

Quit: button that is used to end the review of new taxa without saving any of the data.

Save: button that is used to add the data to an existing attributes file or create a new file. Use this button only after all taxa have been reviewed and updated. IDAS will reset the Edit data module after saving the attributes data.

Taxonomic hierarchy for BU_ID: lists the full taxonomic hierarchy for the BU_ID. Entries in the hierarchy can be transferred to the NAME text box by double clicking on the name (for example, double clicking on "*Hydropsyche*"). The Close button will hide the taxonomic hierarchy frame from view. Clicking on the View hierarchy button will bring back the hierarchy frame back into view.

Explanation: provides instructions on how to review taxa.

The information in the STATUS, BU_ID, NAME, EQTXTOL, and EQTXFG fields should be reviewed to ensure that the NAME corresponds to the BU_ID without authority name and the EQTXTOL and EQTXFG names (and values) are acceptable. If not, edit the NAME and use the Lookup button to determine if a better match can be found. Standard MS Windows® 'cut' and 'paste' methods can be used to extract names from the other fields on this form to minimize spelling errors. However, only the NAME field can be changed either by typing in the NAME field or using the editing buttons (Left, Right) to the right of the name field. For example, in figure 32 the BU_ID contains a provisional identification (*Hydropsyche rossi* Flint, Voshel, and Parker/*simulans* Ross). IDAS automatically populates the NAME field with the BU_ID without authority names (*Hydropsyche rossi/simulans*) and finds tolerance and functional group matches at the genus level. The Left|Right editing buttons can be used to enter *Hydropsyche rossi* or *Hydropsyche simulans* into the NAME field and the Lookup button can be used to check for matches. Figure 32 shows that when NAME is set to *Hydropsyche rossi*, tolerance values are found at the genus level and functional group values at the species level. The Update button is used to save these changes to the temporary recordset (that is, STATUS would change to Modified indicating that the NAME differs from what was originally assigned by IDAS). Once the user is satisfied with all the associations between BU_ID and NAME, the Save button is used to save the data to a new or existing attribute file.

Option 5 (Update the ATTRIB spreadsheet) behaves somewhat differently from the other options for creating or modifying attribute files. The primary function of option 5 is to import tolerance or functional group data from another source such as a State biomonitoring program. This option begins with the selection of the attribute (tolerance or functional group) that is to be modified (fig. 34). The attribute file that is to be modified is then opened followed by the file (fig. 5) and spreadsheet or table (fig. 6) containing the new attribute information. The format of the file supplying the new attribute data must consist of a column named **SEQ** that is populated with a consecutive number identifying the rows of data, a column named **NAME** that is populated with the name of the taxon without authority name, and one or more columns of attributes associated with the taxon (table 24). The column names that identify the attributes must be the same as those used in the ATTRIB spreadsheet (**MATL_TOL, SE_TOL, UMW_TOL, MW_TOL, NW_TOL,** or **FG**). The attribute column names are used to identify the attributes that will be replaced in the ATTRIB column as specified by the selection of the attribute to be modified (fig. 34). The program provides the user with several options for merging taxa, updating attributes, and recalculating the national tolerance values (fig. 35).

The first option for merging taxa (fig. 35) adds taxa and their attributes to the ATTRIB spreadsheet only if the taxa from the external file correspond to taxa (NAME) that already exist in the HIER spreadsheet. The second option adds all of the taxa and attributes that exist in the external source to the ATTRIB spreadsheet. The last option for merging taxa lists only incorporates attributes for taxa that already exist in the ATTRIB spreadsheet; no new taxa are added. The attributes can be updated either by combining or replacing the new attributes with the old attributes. The combining option replaces the information in the ATTRIB spreadsheet only when the taxon in the ATTRIB spreadsheet occurs in the external source, otherwise the value in ATTRIB is retained. The Replace option deletes all the attribute data for all the taxa in ATTRIB and adds the attribute values from the external source. If the taxon does not exist in the external source, the attribute information in ATTRIB is lost for the specified attribute and region. The national tolerance value (NAT_TOL) can also be recalculated as the mean of the revised regional tolerance values or the original NAT_TOL values can be retained. Once the merge taxa, update attributes, and recalculate NAT_TOL options have been selected, clicking on the **Select** button (fig. 35) updates the ATTRIB spreadsheet.

Option 5 (Update the ATTRIB spreadsheet) updates functional group (FG) or tolerance values for a region, but it does not update the EQTX spreadsheet. Options 1–4 must be used to update the EQTX spreadsheet before calculating tolerance or functional group metrics based on data imported from an external source.

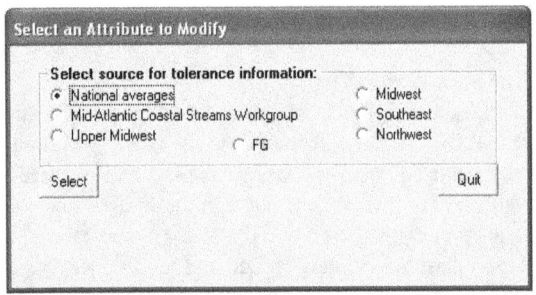

**Figure 34.**    The Select an Attribute to Modify window allows the user to select the attribute that is to be modified in option 5 (Update the ATTRIB spreadsheet).

**Table 24.**    Example of a file holding tolerance and functional group information that can be used to update the attribute file.

| SEQ | NAME | NW_TOL | FG |
|---|---|---|---|
| 1 | *Serratella* | 0.6 | GC |
| 2 | *Serratella deficiens* | 2.1 | |
| 3 | *Ephemera* | 3.1 | GC |
| 4 | *Hexagenia* | 3.6 | GC |
| 5 | *Hexagenia limbata* | 2.6 | GC |
| 6 | *Tricorythodes* | 2.7 | GC |
| 7 | *Choroterpes* | 4.0 | GC |
| 8 | *Paraleptophlebia* | 2.8 | GC |
| 9 | *Anthopotamus* | 3.2 | |
| 10 | *Acerpenna macdunnoughi* | 1.1 | SH |
| 11 | *Acerpenna pygmaea* | 2.3 | OM |
| 12 | *Baetis* | 3.1 | GC |
| 13 | *Baetis flavistriga* | 2.9 | GC |
| 14 | *Baetis intercalaris* | 2.7 | OM |

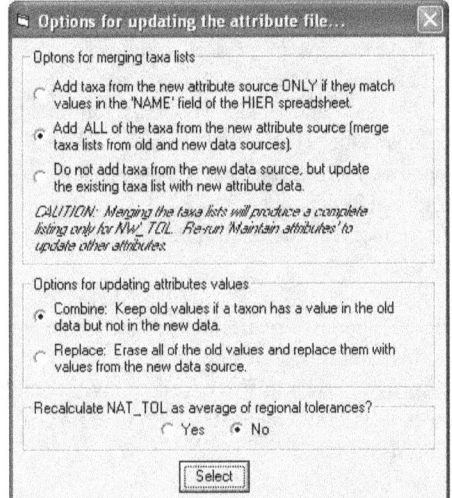

**Figure 35.**    The Options for updating the attribute file window allows the user to determine if new taxa will be added to the ATTRIB spreadsheet, how regional attributes will be updated, and whether national tolerances (NAT_TOL) will be recalculated.

TIP: Updating the ATTRIB spreadsheet does not update the EQTX spreadsheet. Create a new or modify an existing attribute file after updating the ATTRIB spreadsheet to ensure that the EQTX spreadsheet is updated.

## Utilities

The Utilities submenu in the Edit Data module allows the user to convert data files from matrix format to stacked column format (Convert matrix to columns) or to select a random subsample from a dataset (Random subsample). The first option allows the user to convert data in matrix format (taxa as rows and samples as columns or samples as rows and taxa as columns) to a stacked column format. Data in stacked column format can then be imported using the Import Data option. The Random subsample option allows the user to create a new file that contains a random subsample of data from a dataset.

## Convert Matrix to Columns

The Convert matrix to columns function can convert Access and Excel (2003 and 2007) data files in matrix format to the stacked column format that can be imported into IDAS (Import Data). The data matrices can have taxa as rows and samples as columns (Abund_Data_TR.xls and Full_Data_TR.xls) or samples as columns and taxa as rows (Abund_Data_TC.xls and Full_Data_TC.xls). The columns of the data matrices can contain one variable (Abund_Data_TR.xls, Abund_Data_TC.xls, Full_Data_TC.xls) or multiple variables (Full_Data_TR.xls) with different data types (for example, numeric, text, or date/time). Matrices in which columns contain multiple data types require that all data are formatted as text strings because IDAS uses an Access database to perform many of its calculations and Access allows only one data type (text, integer, long, double, date/time) per data column.

Because Excel places no restrictions on the data types that can occupy a column, Excel spreadsheets can be created with multiple data types in a column (Full_Data_TR.xls). The values in this type of Excel file must be converted to text before they can be converted to stacked column format. This cannot be done simply by setting the cell formats to text. Instead, use the "text" function in Excel to convert numeric data (=text(A3,"######0.0000")) or date/time data (=text(A3,"MM/DD/YYYY")) to text before attempting to convert the matrix. Similarly, an Access table that holds a data matrix with multiple data types per column must have the data type of each field set to "text" in order to hold text, numeric, and date/time data.

Of the four taxa-by-sample data matrices that can be converted to stacked format, two (Abund_Data_TR and Abund_Data_TC) do not contain sample information such as collection date, sample type, replicate, and sampling location. This information must be provided in a separate file (Sample_Info.xls) that can be used when importing the data into IDAS (Import Data). The other example files (Full_Data_TR and Full_Data_TC) contain sample information that can be used when importing data after converting it to stacked column format. None of these example files contain information on the taxonomic hierarchy. Consequently, a separate taxonomic hierarchy file (Hier.xls) would have to be supplied when importing these data.

When creating data matrices in Access and Excel, keep in mind that some versions will only allow about 256 columns of data. This limit can easily be exceeded when taxa are represented as columns in the data matrix. Also, be aware that Access data tables will not accept periods (.) as part of a field name (for example, *Hydropsyche* sp.). If a field name contains a period in Excel, it will automatically be converted to a pound sign (for example, *Hydropsyche* sp#, Site abbrev#) when it is transferred to an Access database. This is an issue for data matrices that have taxa as columns or that use periods in field names (for example, Site abbrev.). IDAS automatically detects cases where the # symbol is substituted in taxa names and will convert the "#" symbol back to a period (*Hydropsyche* sp# to *Hydropsyche* sp.). However, IDAS may not correct for this substitution in other field names. Therefore, it is highly recommended that periods not be used in field names other than taxa names.

Selecting the Convert matrix to columns submenu calls up the standard file selection (Excel or Access) and table or spreadsheet selection windows (figs. 5 and 6). Once the data matrix has been opened, the program will display the **Convert Site by Taxa matrix to columns** frame (fig. 36) which provides options for (❶) specifying whether the rows of the matrix represent taxa or samples (that is, taxa as rows and samples as columns or samples as rows and taxa as columns). By default, the program will extract data as if the matrix format was taxa as rows and samples as columns. If the **Data fields** list (❸) displays the contents of a data field rather than a list of field names, then the option for extracting data from the matrix (❶) must be changed. The number of rows and columns in the data matrix are displayed (❷) above the **All non-taxonomic fields** list (❺). The arrow buttons (❹) are used to transfer data between the **Data fields** and **All non-taxonomic fields** list boxes (❸,❺). Multiple items can be transferred by using Ctrl+right mouse button to select multiple items one at a time or Shift+right mouse button to select a continuous range of items. Items can also be transferred by double clicking on an entry, which will transfer it to the other list box. Data are prepared for conversion by transferring any non-taxonomic information from the **Data fields** list box (❸) to **All non-taxonomic fields** list box (❺). Once the non-taxonomic information has been transferred, clicking on the **Convert** button (❻) will convert the data matrix to the

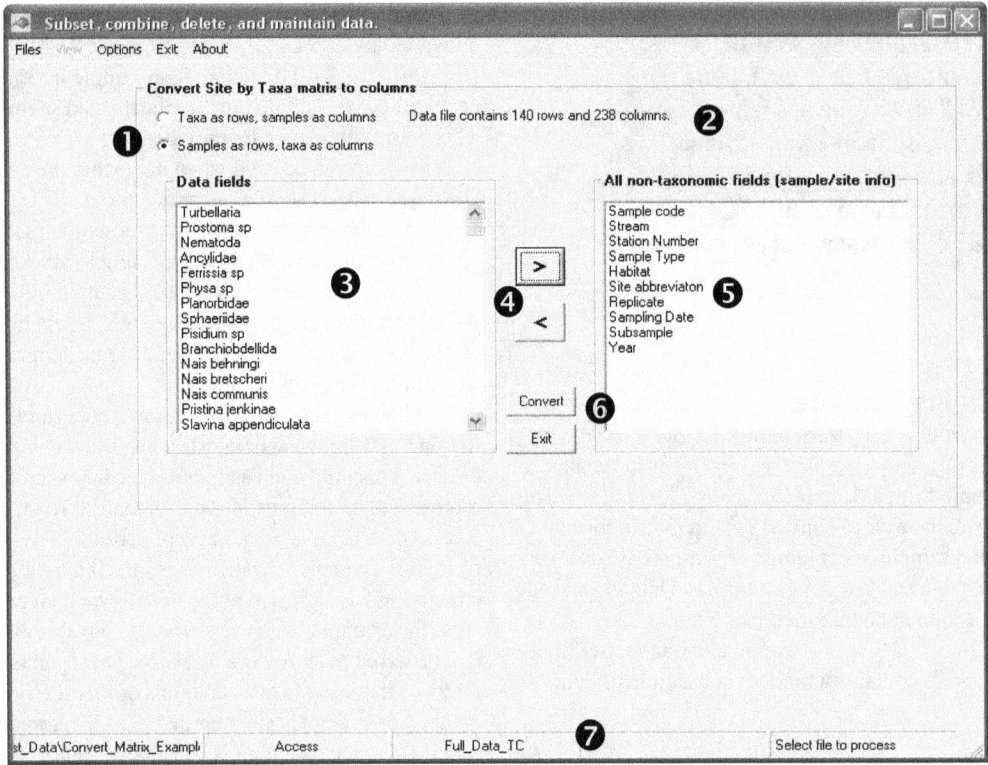

**Figure 36.**    The Convert matrix to columns option is used to convert a data matrix (taxa as rows, samples as columns or samples as rows, taxa as columns) to a stacked column format that can then be imported into IDAS.

[❶ Select the format of the matrix supplying the data, either taxa as rows and samples as columns (default setting) or samples as rows and taxa as columns. ❷ Number of rows and columns in the data matrix. ❸ Data fields in the data matrix. ❹ Transfer buttons used to move data between the list boxes. ❺ Non-taxonomic fields that were selected from the Data fields list. ❻ Buttons used to convert the data to stacked column format or exit this function. ❼ Status bar that indicates the file that has been opened, its type (Excel or Access), the name of the data table or spreadsheet being processed, and the status of data processing]

stacked column format (that is, sample data stacked one on top of another with each row representing a taxon in the sample). This format can then be converted to Bio-TDB format using the User defined formats option under the Data import submenu. The program will request a table or spreadsheet name to use for storing the data (fig. 14). The new data file will be saved in the same Excel workbook or Access database that provided the data matrix.

## Random Subsample

The Random Subsample option extracts a random subsample of data from a dataset and saves it to a new Excel spreadsheet or Access table. The random subsample is based on the number of organisms that compose the subsample. The most common use for this procedure is to simulate a subsample in order to make the data more comparable with another dataset. For example, a 100 count subsample could be drawn from data derived using a 300 count subsample in order to make the data more comparable with other samples derived using a 100 count subsample. This procedure should only be applied to the laboratory counts before they are corrected for laboratory and field subsampling and sampling area.

The random subsample procedure is accessed from the Utilities submenu. Selecting the Random subsample submenu brings up the standard open file and table/spreadsheet windows (figs. 5 and 6). This subsampling procedure only operates on files (Excel or Access) in Bio-TDB format. Once a data file has been opened, the subsampling window is displayed (fig. 37). The desired subsample (integer) is entered in the Subsample (counts) field (❶). The maximum, minimum, and mean abundance of organisms in the dataset are displayed in the Statistics frame (❷) along with the number of samples in the dataset (N). Clicking on the Subsample button (❸) begins the subsampling process. Clicking on the Cancel button (❸) terminates the subsampling process and returns the user to the main window of the Edit Data module.

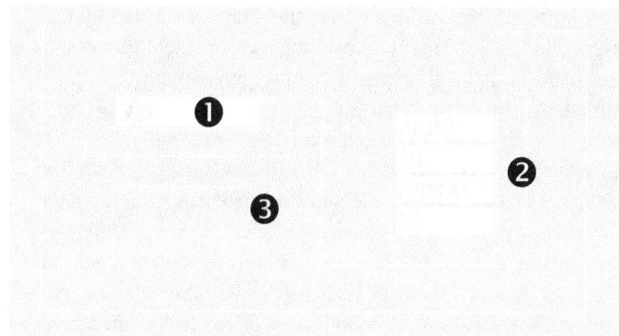

**Figure 37.** The Select a random subset of invertebrates procedure frame showing the subsample specification field (❶), sample statistics fields (❷), and sample processing buttons (❸).

The program creates the subsample by extracting the data for a sample and populating an array with the names of the taxa. The abundance (rounded to the nearest whole number) of each taxon determines the number of times that the name appears in the array (that is, the number of occurrences of the taxon in the array equals its abundance in the sample rounded to the nearest whole number). The program randomizes the array elements by stepping through the array one element at a time. At each element of the array, it randomly selects a number between 1 and the size of the array (that is, the total abundance of the invertebrates in the sample) and switches data (taxon names) between the current array element and the randomly selected element. It then moves to the next element of the array and does the same random repositioning of names in the array. It repeats this randomization 500 times, that is, it goes through the entire array 500 times (total number of random re-ordering is 500 * total abundance).

Once the randomization has been completed, the subsample is extracted by picking the first "X" elements of the array, where "X" is the subsample size (number of counts) entered by the user. For example, if a 100 count subsample was specified, the first 100 elements of the randomized array would be selected. The program then recombines the data, creating an abundance value for each taxon name and incorporating the appropriate taxonomic, site, and sampling information. If the total abundance in a sample is less than or equal to the requested sample size (counts), the entire sample is copied to the new dataset. The program then reads in the next sample, extracts a subsample, and continues until all samples in the dataset have been processed.

# Data Preparation Module

The Data Preparation module prepares data for analysis by the other modules. This module reads abundance data in Bio-TDB format (table 1) and produces a new file format (processed format, table 3) that can be read by the other IDAS modules. There are two main objectives in data preparation: (1) resolving taxonomic ambiguities in the data and (2) combining data so that rows of data correspond to taxa richness (where a taxon is represented by a BU_ID or a combination of BU_ID and lifestage). The Data Preparation module provides the following functionalities:

1. Select sample types (ALL, QUAL, QMH, RTH, DTH) to process.

2. Calculate densities (number per square meter, no./m²) using the sample area information contained in the file "Sample_All" spreadsheet or table exported by Bio-TDB.

3. Delete data based on NWQL BG processing notes such as immature or damaged.

4. Delete data based on lifestages (pupae or non-aquatic adult insects).

5. Combine or retain lifestage information.

6. Form QUAL samples from QMH, RTH, and DTH samples (QMH+RTH, QMH+DTH, QMH+RTH+DTH).

7. Select a lowest taxonomic level (family, tribe, genus) for the dataset.

8. Delete rare taxa based on:

   A. Percentage of sites in the dataset that contain the taxon and (or)

   B. Percentage contribution of the taxon to total abundance in the sample.

9. Resolve taxonomic ambiguities by:

   A. Removing ambiguous taxa at or above a taxonomic level and (or)

   B. Resolving ambiguous taxa separately for each sample or for a combination of samples using one of five methods:

      (1) Remove parent and keep children (RPKC).

      (2) Merge children with parents (MCWP).

      (3) Remove parent or merge children with parent (RPMC).

      (4) Distribute parent among children (DPAC).

      (5) None – retain ambiguous taxa (ORIG).

> **TIP: Data must be processed by the Data Preparation module before they can be processed by the Calculate Community Metrics, Calculate Diversities and Similarities, and Data Export modules.**

The Files, View, Exit, Run, and About menus operate the same as in other modules. To start the process, click Files/Open and select the file and spreadsheet or data table to process (figs. 5 and 6). The IDAS program will check to make sure that the data are in the correct formats and will alert the user with an error message (for example, fig. 38) if it finds a problem. The Status menu option is used to display a separate window that tracks the progress of the IDAS program

as it performs the various processing steps requested by the user. The Options menu allows the user to select the order for displaying the children of an ambiguous parent (fig. 2) alphabetically or by occurrence (that is, descending order based on the number of sites where each taxon is found). Once an appropriate (that is, Bio-TDB format) Excel® spreadsheet or Access® data table is selected for processing and the data format and contents are checked, the IDAS program displays the various options for data preparation (fig. 39) and activates the Run menu. If the user selects Run without selecting any options, the resulting file will have a different format and fewer rows of data than the original file because the Data Preparation module removes, at a minimum, duplicity in BU_IDs (or BU_ID + Lifestage) associated with NWQL BG sample processing notes.

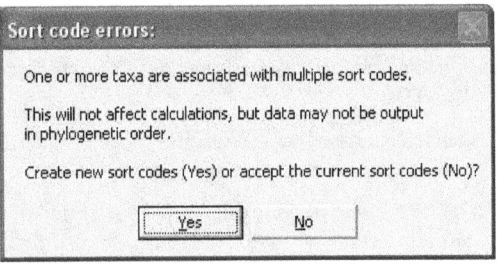

**Figure 38.** Example error message with options for resolving the error in the Data Preparation module.

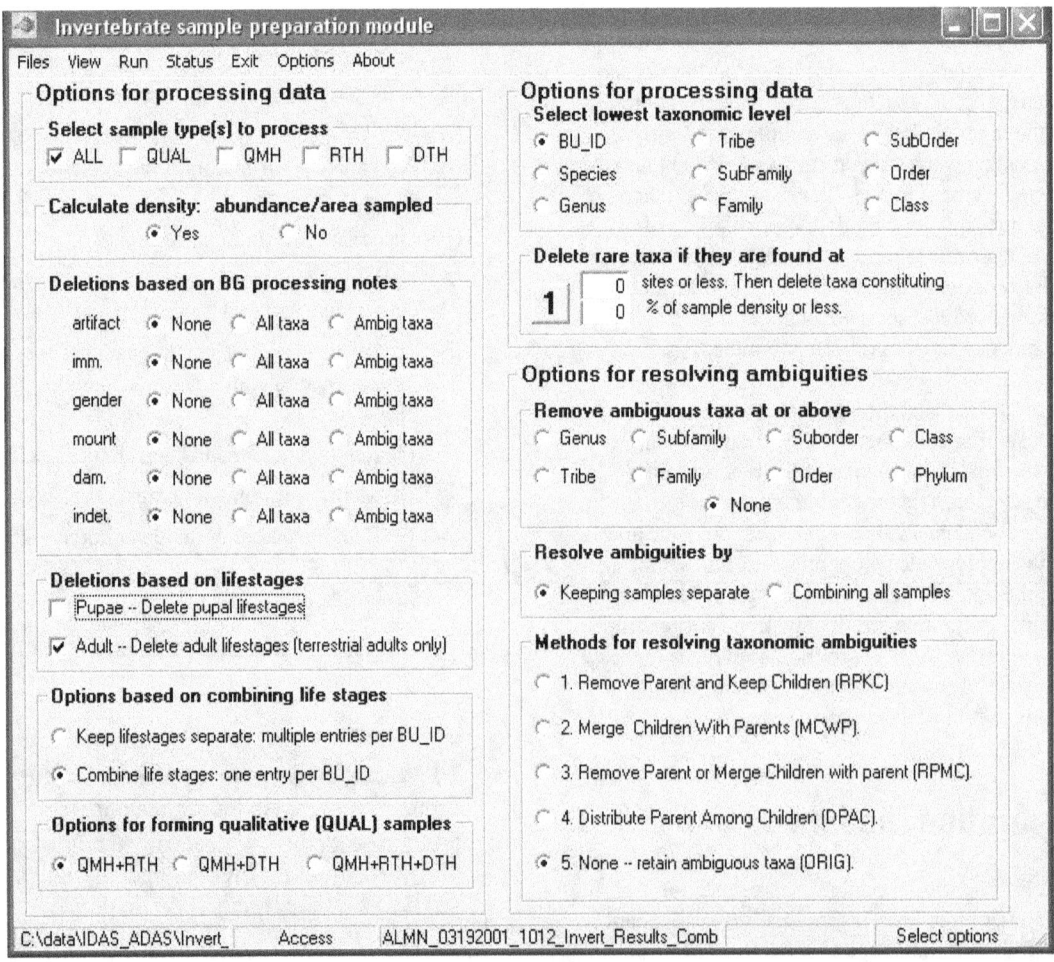

**Figure 39.** Main window of the Data Preparation module with the processing options displayed.

## Processing Options

The Data Preparation module provides a large number of options for processing invertebrate data. These options allow the user to produce datasets that emphasize different characteristics of the data for specific analyses (for example, qualitative lists of taxa associated with a site). Collectively, these options provide a powerful set of tools for manipulating NAWQA Program invertebrate data.

## Select Sample Type(s) to Process

The Select sample type(s) to process option is used to select whether **QMH**, **RTH**, **DTH**, and (or) **QUAL** samples are processed (RWS, Reachwide; SHS, Shore; TRS, targeted riffle; and COMB for WR-EMAP data). QUAL (COMB for WR-EMAP) samples are synthetic samples that are intended to provide a comprehensive taxa list for a site, reach, and date. They are formed by combining QMH, RTH, and DTH samples (RWS, SHS, and TRS for WR-EMAP data) associated with a unique combination of SUID, STAID, REACH, and a range of collection dates (CollectionDate) centered on the QMH sample-collection date. If the user elects to create QUAL samples, the IDAS program will prompt the user to specify a date range (within ± 0 to 7 days of the QMH sample-collection date) over which it will search for samples to combine (fig. 40) with QMH samples. RTH and DTH samples are paired with QMH samples if they have the same SUID, STAID, REACH, and if their collection dates fall within the range specified by the user. If the user elects to create COMB samples for datasets in WR-EMAP format, the COMB samples are formed by combining all the RWS, SHS, and (or) TRS samples collected at a site in the same year. There is no provision to combine WR-EMAP samples collected on a specific date or for a range of dates.

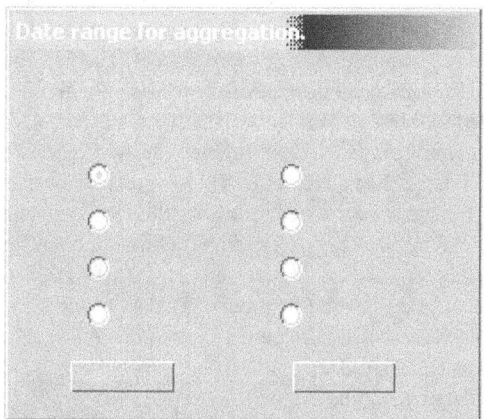

**Figure 40.** Form for selecting the range of dates over which to aggregate RTH, DTH, and QMH samples when creating the QUAL sample.

The date range for QUAL aggregation addresses situations where samples at a site are collected over multiple days, for example, RTH and DTH collected on Friday and QMH on Monday. If the IDAS program finds multiple RTH or DTH samples that can be matched with the QMH sample, it will display a screen that allows the user to select which RTH and (or) DTH samples are to be paired (fig. 41) with the QMH sample. Each list box also contains a blank line at the bottom that can be used to remove a selection. Only one RTH and (or) DTH sample can be paired with each QMH sample. Selecting the Cancel button terminates sample processing and returns to the opening screen of the Data Preparation module. The Accept selection button associates the QMH sample with the selected RTH and DTH samples. The IDAS program automatically documents which RTH and DTH samples are associated with each QMH sample during the creation of the QUAL samples. This documentation consists of the following columns of information that are stored in a spreadsheet or data table with the suffix "_QQSMCODs" (QUAL_QQSMCODs, Appendix II):

**SUID:** four-character abbreviation for Study Unit, text.

**STAID:** station number, text.

**Reach:** sampling reach, text.

**CollectionDate:** date that the QMH sample was collected, date/time.

**QSMCOD:** the SMCOD used to identify the QUAL sample. This SMCOD is generated by setting the "M" identifier in a QMH SMCOD to "Q" (for example, IQM becomes IQQ), text.

**QsampleID:** sampleID for the QUAL sample. This is generated by negating the SampleID assigned to the QMH sample (for example, 7770 becomes -7770), long integer.

**QMHsmcod:** SMCOD assigned to the QMH sample, text.

**QMHSampleID:** SampleID assigned to the QMH sample, long integer.

**RTHsmcod:** SMCOD assigned to the RTH sample, text.

**RTHSampleID:** sampleID assigned to the RTH sample, long integer.

**DTHsmcod:** SMCOD assigned to the DTH sample, text.

**DTHSampleID:** sampleID assigned to the DTH sample, long integer.

> **TIP:** Do not combine qualitative (QMH, QUAL) and quantitative (RTH, DTH) samples in the same data preparation run. Applying quantitative criteria to qualitative data can have unanticipated effects on the resulting data. Separate data files should be created for qualitative and quantitative samples. IDAS is designed to do this in the Data Preparation module without having to re-open the source file.

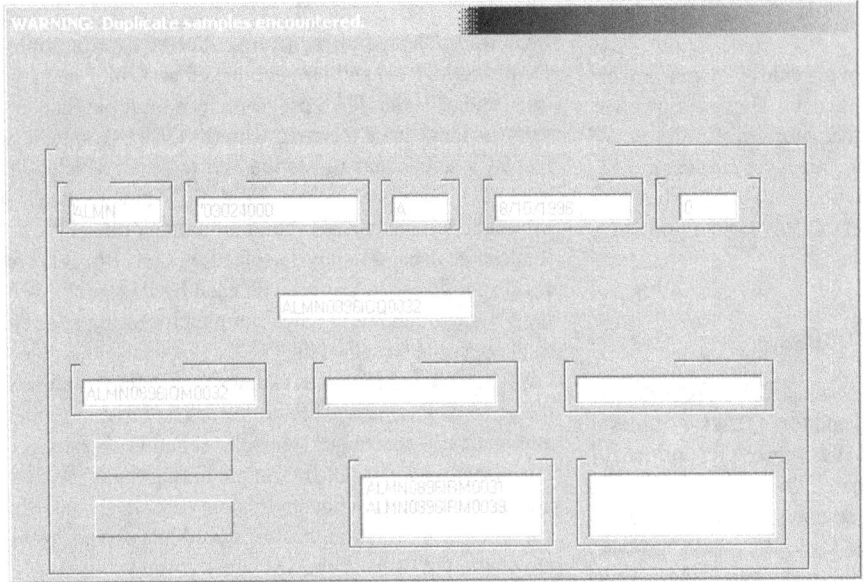

**Figure 41.**   Form for selecting the appropriate RTH and (or) DTH sample to add to the QMH samples when creating the QUAL sample.

TIP: The list boxes associated with "RTH choices" and "DTH choices" include a blank line at the bottom that can be selected to clear the entries in the "RTH SMCOD" or "DTH SMCOD" text boxes.

An equivalent spreadsheet or table of QQSMCODs is created when WR-EMAP format data are processed; however, the column identifiers and sample codes for the combined (COMB) samples are based on the three WR-EMAP sample types (RWS, SHS, TRS) that are combined to form the COMB sample. For example, the table or spreadsheet that contains information on which samples are combined is in the form " _CCSMCODs" (EMAP_CCSMCODs) and the Reach, CollectionDate, SampleID, and SMCOD for the COMB samples are identified as CC_Reach, CC_CollectionDate, CC_SampleID, and CC_SMCOD. CC_SMCODs are derived by combining all samples collected at a site for a year, rather than for a date or range of dates. The SUID portion of the CC_SMCOD is set to "EMAP," month (MM) and year (YY) are extracted from the sample with the earliest collection date, the sample is designated as "ICC," and the 4-digit sample number is set to the smallest SMCOD number associated with the samples combined to form the COMB sample (EMAP0601ICC0123). The CC_SampleID is formed by taking the negative value of the largest SampleID associated with the samples that were combined to form the COMB sample. The samples that were combined to form the COMB sample are listed in the Reach, CollectionDate, SampleID, and SMCOD columns in rows that have the sample CC_SampleID.

## Calculate Densities

The **Calculate density** option is used to convert abundance (number per sample) data associated with quantitative (RTH and DTH) samples (SHS, RWS, TRS for WR_EMAP data) to density (number per square meter, no./m$^2$) data. This feature is disabled if the user has chosen to process only qualitative data (QMH and (or) QUAL). If the user chooses to calculate densities, the program prompts the user to select a spreadsheet or data table that contains information on the area sampled for each RTH or DTH sample. This information is contained in Bio-TDB export files of the form "_Sample_All. xls" (for example, ALMN_10302001_1052_Sample_ All.xls) or as a data table (Sample_All) in the Access database produced by the IDAS export option in Bio-TDB (IDAS_20090715_1104_tcuffney.mdb). If the program cannot find sample area information for a sample, it alerts the user and provides the option of continuing without the sample or quitting the analysis and returning to the opening screen of the Data Preparation module. The user must check to make sure that the "Sample_All.xls" spreadsheet or table contains accurate sample area information; otherwise the densities generated will be in error. If the data that are being processed did not come from Bio-TDB (for example, WR-EMAP or imported data), then a "Sample_All" spreadsheet or data table will have to be created following the format specified in table 2. Sample areas (AreaSampTot) should be in square centimeters (cm$^2$). If qualitative samples (QMH or QUAL) are present in the data being processed, the densities for these samples will automatically be set to one (1).

Densities should be calculated when the sampling areas associated with the samples in the dataset are variable and (or) analyses include comparing abundances among samples. If all sampling areas are the

TIP: Sample area information is located in the Sample_All.xls file in Bio-TDB.

same for all samples (for example, composites of five Slack samplers) or the analytical techniques to be used are strictly qualitative, then densities do not need to be calculated.

## Data Deletions Based on BG Processing Notes

The Deletions based on BG processing notes option allows the user to delete rows of data based on the NWQL BG sample-processing notes (**Notes** column, table 1) associated with each row. The IDAS program recognizes six standard sample-processing notes (Moulton and others, 2000) that can be combined and used to delete data (table 25). One of three processing options (None, All taxa, and Ambig taxa) can be selected for each type of laboratory sample-processing note. None is the default option, and selecting this option results in no deletions. The All taxa option will delete all lines of data where the sample-processing note appears. The Ambig taxa option will only delete a line of data if the note is present and the taxon is an ambiguous parent within the sample. In table 26 the various options are shown for the sample-processing note "imm." (immature). The option None does

not alter the data; the All taxa option deletes all data with the string "imm." in the processing notes column; Ambig taxa deletes only data with "imm." noted in association with an ambiguous parent (Hydropsychidae). This option does not delete data when "imm." is associated with a child (*Hydropsyche*) of an ambiguous parent-child pair.

Processing notes provide a means for the NWQL BG taxonomists to communicate problems with data quality to the analyst. The analyst can then decide whether to eliminate poor-quality data (for example, damaged specimens) on the basis of these notes. Eliminating damaged (dam.) and immature (imm.) specimens can be an effective method of reducing taxonomic ambiguities in the data as seen in table 27, which compares the data before (table 27A) and after (table 27B) eliminating immature (imm.) specimens. Using the Ambig taxa option of this function will minimize the loss of taxa richness in samples by restricting the elimination of taxa to those instances when the taxa are ambiguous. The decision on whether to delete taxa based on processing notes depends on the amount of information (taxa richness and abundance) that will be lost in relation to the reduction in ambiguous taxa.

**Table 25.** National Water Quality Laboratory Biological Group (NWQL BG) standardized sample-processing notes (Moulton and others, 2000) that are recognized by the IDAS program.

[The user can delete rows of data based on any combination of these notes]

| Notes | Description |
|---|---|
| Artifact | Artifact: taxon is not represented by a complete organism (for example, bryozoan fragment or empty mollusk shell). |
| imm. | Immature: identification to a prescribed level is not supported because the organism(s) is(are) too immature. |
| Gender | Gender: identification to targeted level is not supported because of gender. |
| Mount | Poor mount: identification to a prescribed level is not supported because slide-mounted organism(s) is(are) poorly oriented on slide. |
| dam. | Damaged: identification to a prescribed level is not supported because the organism(s) is(are) damaged. |
| indet. | Indeterminate: identification to a prescribed level is not supported for recently molted organisms, mayfly subimagos, or mature and intact organisms because of undocumented variation or indistinct characters, required case is missing or damaged, or required habitat or ecological information is missing or unavailable. |

**Table 26.** The three options for applying NWQL BG sample-processing notes operate differently if ambiguous taxa are present. In this example, Hydropsychidae is an ambiguous parent of *Hydropsyche* and *Ceratopsyche*.

[imm., immature; dam , damaged]

| BU_ID | Abundance | Notes | Abundances after processing notes using option: | | |
|---|---|---|---|---|---|
| | | | None | All taxa | Ambig taxa |
| Hydropsychidae | 10 | imm. | 10 | 0 | 0 |
| *Hydropsyche* | 20 | dam., imm. | 20 | 0 | 20 |
| *Ceratopsyche* | 30 | | 30 | 30 | 30 |

**Table 27.**   Example of how the IDAS program can modify data using sample-processing notes (Notes) and lifestage data. How the data are combined affects the number of rows in the data (richness) and the total abundance in the processed sample.

[L, larva; P, pupa; A, adult; ref., reference; imm., immature; dam., damaged]

| BU_ID | Lifestage | Notes | Abundance |
|---|---|---|---|
| A. Data prior to processing: | | | |
| *Hydroptila* sp. | L | ref. | 1 |
| *Hydroptila* sp. | P | | 5 |
| *Hydroptila* sp. | A | | 5 |
| *Hydroptila* sp. | L | imm. | 50 |
| *Hydroptila* sp. | L | dam. | 62 |
| *Hydroptila* sp. | L | dam., imm. | 65 |
| *Optioservus* sp. | A | | 20 |
| *Optioservus* sp. | L | | 40 |
| | | Taxa richness | 8 |
| | | Total | 248 |
| B. Data after eliminating organisms too immature (imm.) to identify (using option "All taxa"): | | | |
| *Hydroptila* sp. | L | ref. | 1 |
| *Hydroptila* sp. | P | | 5 |
| *Hydroptila* sp. | A | | 5 |
| *Hydroptila* sp. | L | dam. | 62 |
| *Optioservus* sp. | A | | 20 |
| *Optioservus* sp. | L | | 40 |
| | | Taxa richness | 6 |
| | | Total | 133 |
| C. Data after eliminating immature (all) and non-aquatic adult insects: | | | |
| *Hydroptila* sp. | L | ref. | 1 |
| *Hydroptila* sp. | P | | 5 |
| *Hydroptila* sp. | L | dam. | 62 |
| *Optioservus* sp. | A | | 20 |
| *Optioservus* sp. | L | | 40 |
| | | Taxa richness | 5 |
| | | Total | 128 |
| D. Data after eliminating immature (all), non-aquatic adult insects, and combining lifestages: | | | |
| *Hydroptila* sp. | | | 68 |
| *Optioservus* sp. | | | 60 |
| | | Taxa richness | 2 |
| | | Total | 128 |
| E. Data after eliminating immature (all), non-aquatic adult insects, and keeping lifestages separate: | | | |
| *Hydroptila* sp. | L | | 63 |
| *Hydroptila* sp. | P | | 5 |
| *Optioservus* sp. | A | | 20 |
| *Optioservus* sp. | L | | 40 |
| | | Taxa richness | 4 |
| | | Total | 128 |

## Deletions Based on Lifestages

The Deletions based on lifestages option allows the user to delete all lines of data with a lifestage listed as "P" for pupa and (or) data that correspond to non-aquatic adult insects. Non-aquatic adult insects are defined as adults in the orders Ephemeroptera, Odonata, Plecoptera, Megaloptera, Neuroptera, Trichoptera, Lepidoptera, Hymenoptera, and Diptera. Adults in the orders Orthoptera, Hemiptera, and Coleoptera will be retained even if the user selects the Delete adult lifestages (terrestrial adults only) option. The program default is to retain pupae and drop non-aquatic adult insects. Table 27C is an example of how this function works in relation to deletions based on BG notes and lifestages. Adults of the caddisfly *Hydroptila* are eliminated, but not adults of the beetle *Optioservus*.

The methods for collecting invertebrate samples in the NAWQA Program (Cuffney and others, 1993; Moulton and others, 2002) are not designed to collect aquatic insects during the terrestrial stage of their lifecycles. Any such adults that are collected are not of much value in community assessments because their place of origin is uncertain. Therefore, it is recommended that terrestrial adults be deleted, which is the default in the IDAS program. Pupae should be retained for analysis because they represent the completion of the aquatic phase of the lifestage and are a part of the invertebrate community at the site.

## Options Based on Combining Lifestages

The choices under Options based on combining lifestages allow the user to keep lifestages (A = adult, P = pupa, L = larva) separate or to combine lifestages. If the user elects to keep lifestages separate, then a taxon is identified by the combination of BU_ID and lifestage, and the processed data can have multiple BU_IDs within a sample (for example, *Lara* L, *Lara* A, and *Lara* P). If the user elects to combine lifestages, there only will be one BU_ID within a sample, and information on lifestage will be lost (for example, *Lara*). It is important for the user to realize that this option will always combine data within a sample based either on BU_ID or BU_ID+lifestage. The resulting datasets usually will have fewer rows than the original data because the data are combined without regard to the NWQL BG sample-processing notes. Combining data on the basis of BU_ID or BU_ID+lifestage requires combining laboratory notes, which renders this information useless as a means of identifying data-quality problems. Consequently, laboratory notes are processed prior to combining lifestages.

Options for handling laboratory processing notes and lifestages can have important consequences in the calculation of metrics by later modules that assume each row in a sample corresponds to a unique taxon. The effects of these options

on estimates of abundance and richness are illustrated in table 27. By eliminating immature organisms, total abundance is reduced from 248 (table 27A) to 133 (table 27B) and the number of lines of data (richness) from 8 (table 27A) to 6 (table 27B). An example of the effects of deleting non-aquatic adult insects is given in table 27C in which adults of the caddisfly *Hydroptila* were deleted but adults of the beetle *Optioservus*, which are aquatic adults, were not. The number of rows (richness) was reduced from 6 to 5 and total abundance changed from 133 to 128. In table 27D the results of combining lifestages are a reduction in the number of lines of data (richness) from 5 to 2 but no change in total abundance. The effects of retaining lifestages can be seen in table 27E in which the number of lines of data (richness) was reduced from 5 to 4, but total abundance remained the same as in table 27C. Because these processing options can have substantial effect on richness and abundance, the user must carefully consider how the options selected can affect the processed data.

## Options for Forming Qualitative (QUAL) Samples

The choices under Options for forming qualitative (QUAL) samples allow the user to specify the sample types to be used in forming the QUAL sample. The user can elect to combine all sample types QMH+RTH+DTH, or RTH+QMH, or DTH+QMH. Even though DTH samples are no longer collected in the NAWQA Program (Moulton and others, 2002) some DTH samples may be present in earlier datasets (Cuffney and others, 1993). For data derived from WR-EMAP data files, the choices for forming the QUAL (COMB) sample are Shore, Targeted Riffle, and (or) Reachwide. This option provides the user with a simple means of controlling the types of data that are used in forming QUAL samples. The default option (RTH+QMH) ensures comparability in the creation of QUAL samples for NAWQA Program datasets collected during the first and second cycles of the Program.

The IDAS program automatically creates new SampleIDs and SMCODs for the QUAL samples based on the QMH samples that form the basis of the QUAL samples. **QUAL SampleIDs** are set to the negative value of the QMH SampleID (for example, 7909 becomes –7909). The sample component designator in the QMH SMCOD ("M") is set to "Q" for the QUAL SMCOD (for example, ALM-N0695IQM0063 becomes ALMN0695IQQ0063). In this way, it is easy to associate a QUAL sample with the QMH sample from which it is derived. The IDAS program documents the samples that were combined to create the QUAL sample and the SampleID and SMCOD created for the QUAL sample. This information is stored in a spreadsheet or table with the suffix "_QQSMCODs." The same type of information is created for the COMB samples that are created when processing data in the WR-EMAP format. This information is stored in a spreadsheet or table with the suffix "_CCSMCODs."

## Options for Processing Data

Options for processing data include the ability to select the lowest taxonomic level in the dataset and the ability to remove rare taxa from the dataset.

### Select Lowest Taxonomic Level

The Select lowest taxonomic level option allows the user to select the lowest taxonomic level that occurs in the dataset (phylum being the highest and species being the lowest). This option works by substituting the selected taxonomic level for lower taxonomic levels in the BU_ID column. Table 28 is an example of how this option operates when the lowest taxonomic level is set to species, genus, family, and order. Once the substitutions are made, the program recombines data on the basis of SampleID, revised BU_ID, and lifestage. The taxonomic hierarchy is then updated to reflect the revised BU_IDs. The IDAS program generates a SortCode for each of the revised BU_IDs when one is not already available. This new SortCode is the highest value of the SortCodes associated with the original BU_IDs that were combined into the new BU_ID. These SortCodes permit the sorting of the data into phylogenetic order, but they may not correspond to the SortCodes in the original dataset.

When selecting the lowest taxonomic level for processing, it is important to note that there can be a difference between datasets produced with the lowest taxonomic level set to BU_ID and datasets produced with the lowest taxonomic level set to species. It would seem that using either of these levels would result in identical datasets, because species is the lowest taxonomic level supported by the IDAS program. However, the BU_ID provided by the NWQL BG can contain provisional and conditional identifications. Provisional and conditional identifications (Moulton and others, 2000) occur when a specimen represents a potentially undescribed species or possesses sufficient characteristics to permit its assignment to two or more closely related species or genera. Provisional and conditional identifications are reported in the BU_ID

column of the data exported from Bio-TDB but are not listed in the taxonomic hierarchy associated with the BU_ID (table 5), which lists only definitive identifications. Consequently, when species is selected as the lowest taxonomic level, the resulting dataset can be different from that based on BU_ID. For example, in table 28 the BU_ID column contains two provisional taxa: the species *Hydropsyche betteni/ depravata* and the genus *Bezzia/Palomyia*. When species are substituted for BU_IDs, the provisional and conditional identifications are replaced with definitive identifications at the next highest taxonomic levels, genus (*Hydropsyche*) and subfamily (Ceratopogoninae), respectively. If the analyst wants to include provisional identifications in analyses, BU_ID must be selected as the lowest taxonomic level. If the analyst does not want to include provisional identifications in the analyses, then species can be selected as the lowest taxonomic level. The user can select different lowest taxonomic levels for different taxonomic groups by using the subset by taxonomy function of the Edit Data module to create subsets of data that can be processed by the Data Preparation module and recombined using the Combine data function of the Edit Data module. Greater control of the representation of provisional and conditional taxa can be achieved by using the Resolve conditional/ provisional taxa function in the Edit Data module prior to running data through the Data Preparation module.

### Delete Rare Taxa

The Delete rare taxa option allows the user to delete taxa (based either on BU_ID or BU_ID + lifestage) that occur at only a few sites and (or) that constitute only a small proportion of the abundance in a sample. The concept of deleting rare taxa prior to analysis is somewhat controversial. Users should review the works of Goff (1975), Faith and Norris (1989), Marchant (1990), Marchant and others (1997), Cao and Williams (1999), and Cao and others (1998; 2001) to obtain some insight into how removing rare taxa can affect multivariate and metric-based analyses.

**Table 28.** An example of how the Select lowest taxonomic level option operates at the species, genus, family, and order levels.

| Original BU_ID | Revised BU_ID when lowest taxonomic level is set at: | | | |
| --- | --- | --- | --- | --- |
| | Species | Genus | Family | Order |
| Glossomatidae | Glossomatidae | Glossomatidae | Glossomatidae | Trichoptera |
| *Agapetus* | *Agapetus* | *Agapetus* | Glossomatidae | Trichoptera |
| *Glossosoma* | *Glossosoma* | *Glossosoma* | Glossomatidae | Trichoptera |
| Hydropsychidae | Hydropsychidae | Hydropsychidae | Hydropsychidae | Trichoptera |
| *Hydropsyche* | *Hydropsyche* | *Hydropsyche* | Hydropsychidae | Trichoptera |
| *H. betteni* | *H. betteni* | *Hydropsyche* | Hydropsychidae | Trichoptera |
| *H. betteni/depravata* | *Hydropsyche* | *Hydropsyche* | Hydropsychidae | Trichoptera |
| *Ceratopsyche* | *Ceratopsyche* | *Ceratopsyche* | Hydropsychidae | Trichoptera |
| *Bezzia/Palomyia* | Ceratopogoninae | Ceratopogoninae | Ceratopogonidae | Diptera |

Four options are available for removing rare taxa. The user can cycle through these options by clicking on the numbered button in the Delete rare taxa if they are found at frame (fig. 39). Options for deleting rare taxa are based on their occurrence at a certain number of sites (integer) or as a certain percentage (0–100) of abundance or density. The available options for deleting rare taxa are:

1. **Delete taxa that occur at less than or equal to the specified number of sites in the dataset. Then delete taxa that constitute less than or equal to a specified percentage of abundance or density in the sample (delete sites, then abundance).** This approach evaluates each element separately and sequentially. That is, the number of sites where each taxon occurs is calculated and the taxa that occur at or below the criterion are deleted. The IDAS program then calculates the percentage abundance or density for each of the remaining taxa in each sample and deletes those that occur at or below the criterion.

2. **Delete taxa that constitute less than or equal to the specified percentage of abundance or density in each sample. Then delete taxa that occur at less than or equal to a specified number of sites in the dataset (delete abundance, then sites).** As in Option 1, this approach evaluates each element separately and sequentially. That is, the program calculates the percentage of abundance or density contributed by each taxon in each sample and deletes those that occur at or below the specified criterion. The program then calculates the number of sites where each of the remaining taxa occurs and deletes the taxa that occur at or below this criterion.

3. **Simultaneously delete taxa that occur at less than or equal to the specified number of sites <u>AND</u> that constitute less than or equal to the specified percentage of abundance (delete sites and abundance).** This option differs from options 1 and 2 in that the number of sites where each taxon occurs and the contribution of each taxon to total abundance/density in each sample are calculated simultaneously. Then these criteria are applied in a simple combined query, that is, a taxon is deleted if the number of sites where it occurs is at or below the criterion level <u>and</u> if the percentage of abundance or density contributed by the taxon in the sample is less than or equal to the criterion. Both criteria must be met before a taxon can be deleted from a sample.

**WARNING: The delete rare taxa options provided in the IDAS program are powerful and can be complicated. Be sure to study the examples provided and understand how each option works before selecting an option to apply to the data! Different options can result in deleting widely different groups of taxa.**

4. **Simultaneously delete taxa that occur at less than or equal to the specified number of sites (delete abundance or sites) <u>OR</u> that constitute less than or equal to the specified percentage of abundance (sites or abundance).** This option is similar to option 3, except that taxa are deleted if they occur at less than or equal to the criteria for number of sites <u>or</u> they contribute less than or equal to the criterion for the percentage of abundance or density. Taxa are deleted from a sample if they meet either of these criteria.

The user must be aware of the nuances of these procedures before applying them to the data. For example, if a taxon is deleted on the basis of the number of sites where it occurs (note: this is the number of sites (STAID) where the taxon is found not the number of samples (SampleID) in which it is found), then the taxon will be deleted across all samples. However, if a taxon is deleted on the basis of its contribution to the total sample abundance or density, it will be eliminated only from those samples where it fails to meet the criterion rather than from all samples. Consequently, the order in which deletions are applied in options 1 and 2 (sites then abundance or abundance then sites) can have a profound effect on the taxa that are deleted because the calculations of occurrence and abundance are conducted sequentially. Similarly, the operators <u>and</u> and <u>or</u> used in options 3 and 4, respectively, can

**TIP: Insight into the appropriate values to enter in the "delete rare taxa" criteria boxes can be obtained by using the "Summarize taxa" option in the Edit Data module. This will give a good summarization of the distribution of taxa across sites and samples in the dataset.**

greatly affect the number of taxa that are deleted and the abundance or density that remains in the processed dataset. Examples of these effects are shown in table 29, in which taxa richness and density are shown before and after applying the four options for deleting rare taxa to the example data file NAWQA_03192001_1012_Invert_Results_Comb.xls. The same criteria (delete taxa that occur at two sites or less; delete taxa that constitute 1 percent or less of sample abundance) were used in each of the four options. As indicated in table 29, the user's selection of a deletion option can have a profound effect on taxa richness and abundance when deletions are

based on a combination of abundance and occurrence. However, if taxa are deleted on the basis of a **single** criterion (either occurrence or abundance) then the results obtained using the four options will not differ.

The following series of examples demonstrate how each of the four deletion options process information on occurrence and abundance or density to identify which taxa to delete. Each example is applied to the hypothetical invertebrate density data presented in table 30 and is based on deleting taxa that occur at two or fewer sites and (or) contribute 5 percent or less of sample abundance.

**Table 29.** Effects of "delete rare taxa" options 1–4 on taxa richness and density.

[Rare taxa were deleted if they occurred at five sites or less and (or) made up 1 percent or less of sample abundance. Data were derived from the example data file NAWQA_03192001_1012_Invert_Results_Comb.xls. Program settings were: RTH samples, calculate density, no deletions based on BG processing notes, delete adults, combine lifestages, BU_ID as the lowest taxonomic level, do not resolve ambiguous taxa. Deletions are based on the data in table 30]

| SampleID | Original | Method used to delete rare taxa | | | |
|---|---|---|---|---|---|
| | | Option 1 | Option 2 | Option 3 | Option 4 |
| Taxa richness | | | | | |
| 7909 | 74 | 22 | 20 | 20 | 22 |
| 7650 | 90 | 26 | 16 | 16 | 22 |
| 7799 | 77 | 25 | 19 | 19 | 24 |
| 7856 | 67 | 28 | 18 | 18 | 28 |
| 7719 | 82 | 19 | 15 | 15 | 19 |
| Density | | | | | |
| 7909 | 6,762.4 | 5,440.0 | 5,209.6 | 5,209.6 | 5,440.0 |
| 7650 | 2,032.8 | 1,376.8 | 888.8 | 888.8 | 1,304.8 |
| 7799 | 1,160.0 | 887.2 | 727.2 | 727.2 | 876.0 |
| 7856 | 2,603.2 | 2,130.4 | 1,407.2 | 1,407.2 | 2,130.4 |
| 7719 | 8,629.6 | 7,148.0 | 6,488.0 | 6,488.0 | 7,148.0 |

**Table 30.** Hypothetical density data used to illustrate how the options for deleting rare taxa determine which taxa to delete.

[Occurrence is the number of sites where the taxon is found]

| Taxon | Site 1 | Site 2 | Site 3 | Site 4 | Site 5 | Occurrence |
|---|---|---|---|---|---|---|
| Spp 1 | 107.2 | | | | | 1 |
| Spp 2 | | 26.4 | | | | 1 |
| Spp 3 | 860.8 | 63.2 | 60.8 | 120.8 | 128.8 | 5 |
| Spp 4 | 476.0 | | 19.2 | | | 2 |
| Spp 5 | | | | | 218.4 | 1 |
| Spp 6 | | 26.4 | 16.0 | 45.3 | | 3 |
| Spp 7 | | 23.2 | | | | 1 |
| Spp 8 | 5.0 | | 51.2 | 10.2 | | 3 |
| Spp 9 | | | | 68.0 | | 1 |
| Spp 10 | | 36.8 | 22.4 | 54.4 | | 3 |
| Total | 1,449.0 | 176.0 | 169.6 | 298.7 | 347.2 | 5 |

**Option 1 (delete sites then abundance)** requires three steps to identify and eliminate rare taxa (table 31). Step 1 eliminates all taxa that occur at two or fewer sites (Spp 1, 2, 4, 5, 7, and 9, table 30). Step 2 calculates the percentage of density contributed by the remaining taxa. Step 3 eliminates all taxa that contribute less than 5 percent of abundance in the sample (Spp 8 in samples from sites 1 and 4). Applying option 1 to the data in table 30 results in the loss of 20–75 percent of taxa richness, but less than 5 percent of sample density.

**Option 2 (delete abundance then sites)** requires three steps to identify and eliminate rare taxa (table 32). Step 1 calculates the percentage of density contributed by each taxon at each site. Step 2 eliminates all taxa that contributed less than 5 percent of the density in each sample (Spp 8 in samples from sites 1 and 4). Step 3 eliminates all of the remaining taxa that occur at two or fewer sites (Spp 1, 2, 4, 5, 7, 8, 9, table 30). Applying option 2 to the data in table 30 resulted in the loss of 40–75 percent of taxa richness and from 27–63 percent of sample density. The dataset produced by option 2 (table 32) is different from that produced by option 1 (table 31) even though the only difference between the two methods is the order in which the criteria for deletion (occurrence and density) are applied.

**Option 3 (delete abundance and sites)** requires only two steps to identify and eliminate rare taxa (table 33). The first step calculates the percentage of density contributed by each taxon in the sample and the number of sites where the taxa occur (occurrence). The second step eliminates taxa that occur at fewer than two sites <u>and</u> contribute 5 percent or less to sample density. The results from applying option 3 to the data in table 30 are very different from those obtained by using options 1 (table 31) or 2 (table 32). No taxa are eliminated because no taxa meet both criteria. Spp 1, 2, 4, 5, 7, and 9 (table 33) have occurrences that are below the occurrence criterion, but their contribution to sample density is above the criterion so they are not deleted. Similarly, Spp 8 at sites 1 and 4 falls below the criterion for contribution to sample density, but Spp 8 occurs at more than two sites so it is not deleted. In this example, applying option 3 to the data in table 30 did not change it, although this would not be the case for other datasets (table 29) and (or) using other criteria.

**Option 4 (delete abundance or sites)** is similar to option 3 in that it requires only two steps to identify and eliminate rare taxa (table 34). The first step calculates the percentage of density contributed by each taxon in the sample and the number of sites where the taxa occur (occurrence). The second step eliminates taxa that occur at fewer than two sites <u>or</u> contribute 5 percent or less to sample density. The difference between option 4 and option 3 is the substitution of the "or" operator for the "and" operator used in option 3. In option 3, a taxon must meet both criteria before it can be deleted, whereas in option 4 a taxon must meet only one of the criteria before it can be deleted. Consequently, more taxa are deleted in option 4 than in option 3. The results of using option 4 also differ substantially from the results obtained by using option 1 (table 31) but are identical to the results obtained by using option 2 (table 32). The comparability of results obtained by using options 2 and 4 is serendipitous. Comparisons of these two options using larger datasets and other criteria show that these options give different results (table 29).

Deleting taxa on the basis of their contribution to sample abundance or density does not have any meaning for qualitative samples in which all abundances are equal—that is, all abundances are one (1). Therefore, if the user is processing only qualitative samples (QMH and (or) QUAL), the option to eliminate taxa on the basis of their contribution to sample abundance or density will be inactive (dimmed). This option is only active if the user is conducting an analysis that involves quantitative samples (RTH or DTH) or a combination of quantitative and qualitative samples. It is highly recommended that the Data Preparation module not be used to process mixed qualitative and quantitative datasets. Each data type should be prepared separately and then, if desired, the user can recombine qualitative and quantitative data into a single spreadsheet or data table by using the Combine tables/ spreadsheets option of the Edit Data module.

> **TIP: Analyze quantitative (RTH and DTH) samples separately from qualitative samples (QMH and QUAL). Use the Edit Data module to create separate qualitative and quantitative datasets.**

**Table 31.**   Examples of the steps used to delete rare taxa in option 1 as applied to the data in table 30.

[Deletion criteria are to delete taxa that occur at two sites or less; then delete taxa that contribute 5 percent or less to sample density]

| Taxon | Site 1 | Site 2 | Site 3 | Site 4 | Site 5 | Occurrence |
|---|---|---|---|---|---|---|
| Step 1: Delete taxa that occur at two sites or less (Spp 1, 2, 4, 5, 7, and 9). | | | | | | |
| Spp 1 | | | | | | 0 |
| Spp 2 | | | | | | 0 |
| Spp 3 | 860.8 | 63.2 | 60.8 | 120.8 | 128.8 | 5 |
| Spp 4 | | | | | | 0 |
| Spp 5 | | | | | | 0 |
| Spp 6 | | 26.4 | 16.0 | 45.3 | | 3 |
| Spp 7 | | | | | | 0 |
| Spp 8 | 5.0 | | 51.2 | 10.2 | | 3 |
| Spp 9 | | | | | | 0 |
| Spp 10 | | 36.8 | 22.4 | 54.4 | | 3 |
| Total | 865.8 | 126.4 | 150.4 | 230.7 | 128.8 | |
| Step 2: Calculate the percentage of total abundance contributed by each taxon after eliminating taxa based on whether they occur at more than two sites. | | | | | | |
| Spp 1 | | | | | | 0 |
| Spp 2 | | | | | | 0 |
| Spp 3 | 99.4 | 50.0 | 40.4 | 52.4 | 100.0 | 5 |
| Spp 4 | | | | | | 0 |
| Spp 5 | | | | | | 0 |
| Spp 6 | | 20.9 | 10.6 | 19.6 | | 3 |
| Spp 7 | | | | | | 0 |
| Spp 8 | 0.6 | | 34.0 | 4.4 | | 3 |
| Spp 9 | | | | | | 0 |
| Spp 10 | | 29.1 | 14.9 | 23.6 | | 3 |
| Total | 100.0 | 100.0 | 100.0 | 100.0 | 100.0 | |
| Step 3: Delete taxa that make up 5 percent or less of sample abundance (Spp 8 at sites 1 and 4). Total is the percentage of original density that remains. | | | | | | |
| Spp 1 | | | | | | 0 |
| Spp 2 | | | | | | 0 |
| Spp 3 | 99.4 | 50.0 | 40.4 | 52.4 | 100.0 | 5 |
| Spp 4 | | | | | | 0 |
| Spp 5 | | | | | | 0 |
| Spp 6 | | 20.9 | 10.6 | 19.6 | | 3 |
| Spp 7 | | | | | | 0 |
| Spp 8 | | | 34.0 | | | 1 |
| Spp 9 | | | | | | 0 |
| Spp 10 | | 29.1 | 14.9 | 23.6 | | 3 |
| Total | 99.4 | 100.0 | 100.0 | 95.6 | 100.0 | |

**Table 32.**   Examples of the steps used to delete rare taxa in option 2 as applied to the data in table 30.

[Deletion criteria are to delete taxa that contribute 5 percent or less to sample density; then delete taxa that occur at two sites or less]

| Taxon | Site 1 | Site 2 | Site 3 | Site 4 | Site 5 | Occurrence |
|---|---|---|---|---|---|---|
| Step 1:  Calculate the percentage of total abundance contributed by each taxon. | | | | | | |
| Spp 1 | 7.4 | | | | | 1 |
| Spp 2 | | 15.0 | | | | 1 |
| Spp 3 | 59.4 | 35.9 | 35.8 | 40.4 | 37.1 | 5 |
| Spp 4 | 32.9 | | 11.3 | | | 2 |
| Spp 5 | | | | | 62.9 | 1 |
| Spp 6 | | 15.0 | 9.4 | 15.2 | | 3 |
| Spp 7 | | 13.2 | | | | 1 |
| Spp 8 | 0.3 | | 30.2 | 3.4 | | 3 |
| Spp 9 | | | | 22.8 | | 1 |
| Spp 10 | | 20.9 | 13.2 | 18.2 | | 3 |
| Total | 100.0 | 100.0 | 100.0 | 100.0 | 100.0 | |
| Step 2:  Delete taxa that make up 5 percent or less of sample abundance (Spp 8 at sites 1 and 4. | | | | | | |
| Spp 1 | 7.4 | | | | | 1 |
| Spp 2 | | 15.0 | | | | 1 |
| Spp 3 | 59.4 | 35.9 | 35.8 | 40.4 | 37.1 | 5 |
| Spp 4 | 32.9 | | 11.3 | | | 2 |
| Spp 5 | | | | | 62.9 | 1 |
| Spp 6 | | 15.0 | 9.4 | 15.2 | | 3 |
| Spp 7 | | 13.2 | | | | 1 |
| Spp 8 | | | 30.2 | | | 1 |
| Spp 9 | | | | 22.8 | | 1 |
| Spp 10 | | 20.9 | 13.2 | 18.2 | | 3 |
| Total | 99.7 | 100.0 | 100.0 | 96.6 | 100.0 | |
| Step 3:  Delete taxa that occur at two sites or less (Spp 1, 2, 4, 5, 7, 8, and 9). | | | | | | |
| Spp 1 | | | | | | 0 |
| Spp 2 | | | | | | 0 |
| Spp 3 | 59.4 | 35.9 | 35.8 | 40.4 | 37.1 | 5 |
| Spp 4 | | | | | | 0 |
| Spp 5 | | | | | | 0 |
| Spp 6 | | 15.0 | 9.4 | 15.2 | | 3 |
| Spp 7 | | | | | | 0 |
| Spp 8 | | | | | | 0 |
| Spp 9 | | | | | | 0 |
| Spp 10 | | 20.9 | 13.2 | 18.2 | | 3 |
| Total | 59.4 | 71.8 | 58.5 | 73.8 | 37.1 | |

**Table 33.** Examples of the steps used to delete rare taxa in option 3 as applied to the data in table 30.

[Deletion criteria are to delete taxa that occur at two or fewer sites _and_ delete taxa that contribute 5 percent or less to sample density]

| Taxon | Site 1 | Site 2 | Site 3 | Site 4 | Site 5 | Occurrence |
|-------|--------|--------|--------|--------|--------|------------|
| Step 1: Calculate the number of sites where each taxon occurs (occurrence) and the percentage of sample abundance contributed by each taxon. | | | | | | |
| Spp 1 | 7.4 | | | | | 1 |
| Spp 2 | | 15.0 | | | | 1 |
| Spp 3 | 59.4 | 35.9 | 35.8 | 40.4 | 37.1 | 5 |
| Spp 4 | 32.9 | | 11.3 | | | 2 |
| Spp 5 | | | | | 62.9 | 1 |
| Spp 6 | | 15.0 | 9.4 | 15.2 | | 3 |
| Spp 7 | | 13.2 | | | | 1 |
| Spp 8 | 0.3 | | 30.2 | 3.4 | | 3 |
| Spp 9 | | | | 22.8 | | 1 |
| Spp 10 | | 20.9 | 13.2 | 18.2 | | 3 |
| Total | 100.0 | 100.0 | 100.0 | 100.0 | 100.0 | 5 |
| Step 2: Delete taxa that occur at two sites or less _and_ that contribute 5 percent or less to abundance in the sample. (No taxa are deleted because no taxa meet both of these criteria.) | | | | | | |
| Spp 1 | 7.4 | | | | | 1 |
| Spp 2 | | 15.0 | | | | 1 |
| Spp 3 | 59.4 | 35.9 | 35.8 | 40.4 | 37.1 | 5 |
| Spp 4 | 32.9 | | 11.3 | | | 2 |
| Spp 5 | | | | | 62.9 | 1 |
| Spp 6 | | 15.0 | 9.4 | 15.2 | | 3 |
| Spp 7 | | 13.2 | | | | 1 |
| Spp 8 | 0.3 | | 30.2 | 3.4 | | 3 |
| Spp 9 | | | | 22.8 | | 1 |
| Spp 10 | | 20.9 | 13.2 | 18.2 | | 3 |
| Total | 100.0 | 100.0 | 100.0 | 100.0 | 100.0 | |

**Table 34.** Examples of the steps used to delete rare taxa in option 4 as applied to the data in table 30.

[Deletion criteria are to delete taxa that occur at two or fewer sites or that contribute 5 percent or less to sample density]

| Taxon | Site 1 | Site 2 | Site 3 | Site 4 | Site 5 | Occurrence |
|---|---|---|---|---|---|---|
| Step 1: Calculate the number of sites where each taxon occurs (occurrence) and the percentage of sample abundance contributed by each taxon. | | | | | | |
| Spp 1 | 7.4 | | | | | 1 |
| Spp 2 | | 15.0 | | | | 1 |
| Spp 3 | 59.4 | 35.9 | 35.8 | 40.4 | 37.1 | 5 |
| Spp 4 | 32.9 | | 11.3 | | | 2 |
| Spp 5 | | | | | 62.9 | 1 |
| Spp 6 | | 15.0 | 9.4 | 15.2 | | 3 |
| Spp 7 | | 13.2 | | | | 1 |
| Spp 8 | 0.3 | | 30.2 | 3.4 | | 3 |
| Spp 9 | | | | 22.8 | | 1 |
| Spp 10 | | 20.9 | 13.2 | 18.2 | | 3 |
| Total | 100.0 | 100.0 | 100.0 | 100.0 | 100.0 | 5 |
| Step 2: Delete taxa that occur at two sites or less or that contribute 5 percent or less to abundance in the sample (Spp 1, 2, 4, 5, 7, 8, and 9 are deleted). | | | | | | |
| Spp 1 | | | | | | 0 |
| Spp 2 | | | | | | 0 |
| Spp 3 | 59.4 | 35.9 | 35.8 | 40.4 | 37.1 | 5 |
| Spp 4 | | | | | | 0 |
| Spp 5 | | | | | | 0 |
| Spp 6 | | 15.0 | 9.4 | 15.2 | | 3 |
| Spp 7 | | | | | | 0 |
| Spp 8 | | | | | | 0 |
| Spp 9 | | | | | | 0 |
| Spp 10 | | 20.9 | 13.2 | 18.2 | | 3 |
| Total | 59.4 | 71.8 | 58.5 | 73.8 | 37.1 | |

## Options for Resolving Ambiguities

The IDAS program provides a variety of options for removing and resolving taxonomic ambiguities. Ambiguous taxa can be removed if they occur at or above a specified taxonomic level (Remove ambiguous taxa at or above) and (or) they can be resolved on a sample-by-sample basis (Keeping samples separate) or for all samples combined (Combining all samples). The Remove ambiguous taxa at or above a specified taxonomic level (fig. 39) identifies all the ambiguous taxa in the dataset and then removes those that are at or above the taxonomic level specified by the user. This method is similar to the Remove ambiguous parents function in the Edit Data module.

The sample-by-sample method of resolving ambiguous taxa (Keeping samples separate) identifies and resolves ambiguities separately for each sample and is used when communities are expected to differ widely based on natural environmental factors (for example, data collected across a large geographic area). The combined-sample (Combining all samples) method combines all the data into one sample, finds the ambiguous taxa in the combined sample, and defines rules for resolving ambiguities in the combined sample. These rules are then used to resolve ambiguous taxa in each sample. This method commonly is used in situations where there is an expectation that the communities would be similar among sites in the absence of anthropogenic effects (that is, studies conducted in areas with relatively uniform environmental settings). When ambiguous taxa are identified on a sample-by-sample basis (Keeping samples separate), there can be substantial differences in the ambiguous taxa among the samples. For example, *Baetis* sp. L is an ambiguous parent in samples 7896 and 7973 (table 35) because several species of *Baetis* are present in these samples. However, Baetis sp. L is not an ambiguous taxon in sample 7650 because this sample does not contain any species of *Baetis*. Similar differences can be seen for *Acentrella* sp. L, *Plauditus* sp. L, *Pseudocloeon* sp. L, and Zygoptera L. In contrast, resolving ambiguities for a combined dataset (Combining all samples) results in a consistent set of ambiguous taxa (Combined samples column, table 35) that are resolved uniformly across all samples. However, this approach can lead to instances where a sample contains an ambiguous parent, but no children of that parent (for example, *Baetis* sp. L, *Pseudocloeon* sp. L, and Zygoptera in sample 7650) are present in the sample. The IDAS program provides the user with several methods for handling these situations.

The combined method is most appropriately used in situations where the occurrences of taxa are expected to be fairly uniform among sites in the absence of anthropogenic effects. An example of such a dataset would be a land-use gradient study where sites are chosen within a limited geographical area and environmental settings are as similar as possible. In such situations, there is an expectation that the same taxa will be found at all sites in the absence of human effects. This makes it possible to extrapolate the identity of missing children based on the composition of other samples in the dataset. In contrast, the sample-by-sample method is most appropriately used when analyzing data across large geographic areas (for example, aggregation of data from multiple NAWQA Program Study Units) or when the invertebrate fauna are expected to differ substantially among sites based on natural factors. In these cases, there is no expectation that a taxon that occurs at one site will occur at another. Therefore, it is not appropriate to extrapolate the occurrence of a missing child at one site based on its occurrence at other sites. Instead, such extrapolations should be limited to the occurrences of children in each sample. The method used to resolve ambiguities can have a profound effect on the abundance and taxa richness in a sample (fig. 42), and these effects must be considered when using the processed data to calculate community metrics or in statistical analyses (Cuffney and others, 2007).

---

**Ambiguous taxon:** A taxon in a dataset in which data are reported at one or more lower or higher taxonomic levels in the taxonomic hierarchy. For example, in a sample that contains data for Hydropsychidae, *Hydropsyche*, and *Hydropsyche sparna*, all three taxa are considered ambiguous.

**Ambiguous parent:** A taxon in a group of ambiguous taxa that occurs at a higher taxonomic level than the other taxa in the group. For example, in a sample that contains data for Hydropsychidae, *Hydropsyche*, and *Hydropsyche sparna*, both Hydropsychidae and *Hydropsyche* are ambiguous parents of *Hydropsyche sparna*, and Hydropsychidae is an ambiguous parent of *Hydropsyche*.

**Ambiguous child:** A taxon that occurs at a lower taxonomic level in a group of ambiguous taxa. For example, in a sample that contains data for Hydropsychidae, *Hydropsyche*, and *Hydropsyche sparna*, *Hydropsyche* and *Hydropsyche sparna* are ambiguous children of the ambiguous parent Hydropsychidae, and *Hydropsyche sparna* is the ambiguous child of the ambiguous parents *Hydropsyche* and Hydropsychidae.

**Table 35.** Hypothetical invertebrate data used to illustrate eight methods for resolving ambiguous taxa.

[These data include lifestage information (A = adult, P = pupa, L = larvae) and the identity of ambiguous taxa (Ambig = Yes) for each sample (**Separate option**) and for all sites combined (**Combined option**). "Occur" is the number of samples where each taxon occurs, and "Abund" is the abundance of the taxon]

| Taxon and Lifestage | Combined samples | | | Samples | | | | | | | |
|---|---|---|---|---|---|---|---|---|---|---|---|
| | | | | 7650 | | 7896 | | 7729 | | 7973 | |
| | Occur | Abund | Ambig | Abund | Ambig | Abund | Ambig | Abund | Ambig | Abund | Ambig |
| Ephemeroptera A | 1 | 2 | | 2 | | | | | | | |
| Ephemeroptera L | 1 | 100 | Yes | 100 | Yes | | | | | | |
| Baetidae L | 2 | 12 | Yes | | | 8 | Yes | 4 | Yes | | |
| *Acentrella* sp. L | 3 | 45 | Yes | 5 | Yes | | | 10 | Yes | 30 | |
| *Acentrella parvula* L | 3 | 71 | | 26 | | 11 | | 34 | | | |
| *Acentrella turbida* L | 2 | 12 | | | | 9 | | 3 | | | |
| *Baetis* sp. L | 3 | 179 | Yes | 35 | | 100 | Yes | | | 44 | Yes |
| *Baetis flavistriga* L | 2 | 60 | | | | 6 | | 54 | | | |
| *Baetis intercalaris* L | 2 | 99 | | | | | | 78 | | 21 | |
| *Baetis pluto* L | 3 | 58 | | | | 12 | | 23 | | 23 | |
| *Baetis tricaudatus* L | 2 | 13 | | | | 2 | | 11 | | | |
| *Plauditus* sp. L | 2 | 61 | Yes | 5 | Yes | 56 | | | | | |
| *Plauditus cestus* L | 2 | 87 | | 34 | | | | | | 53 | |
| *Pseudocloeon* sp. L | 2 | 46 | Yes | 45 | | | | 1 | Yes | | |
| *Pseudocloeon propinquum* L | 2 | 117 | | | | | | 5 | | 112 | |
| Zygoptera L | 2 | 110 | Yes | 10 | | 100 | Yes | | | | |
| *Argia* sp. L | 1 | 8 | | | | 8 | | | | | |
| Hydropsychidae L | 1 | 50 | | 50 | | | | | | | |
| *Diplectrona* sp. P | 2 | 33 | | 10 | | | | 23 | | | |
| Total richness | | | | 11 | | 10 | | 11 | | 6 | |
| Total abundance | | | | 322 | | 312 | | 246 | | 283 | |

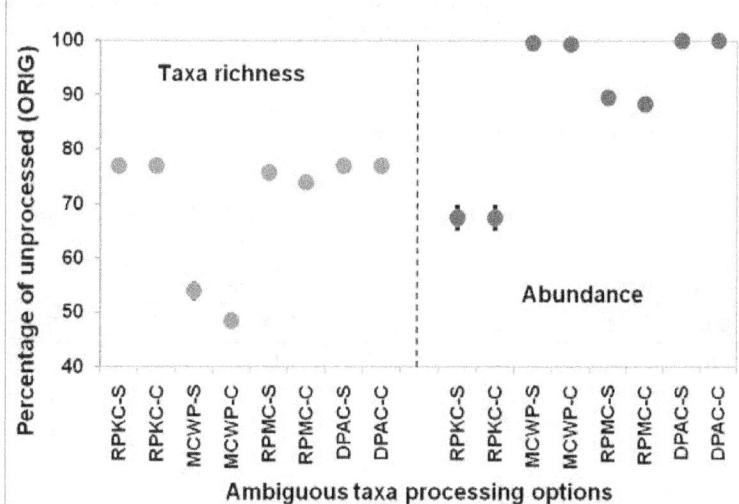

[Results are expressed as a percentage of richness or abundance derived without resolving ambiguities (ORIG) and without taking lifestage into consideration. Means and 95 percent confidence intervals are shown based on RTH samples collected as part of the NAWQA Program urban gradient pilot studies (Cuffney and others, 2007)]

**Figure 42.** The method selected to resolve ambiguous taxa can have a profound effect on taxa richness and abundance. In this example both taxa richness and total abundance are strongly affected by the choice of processing option and whether ambiguities are resolved for each sample separately (-S) or for all samples combined (-C).

Methods for resolving taxonomic ambiguities are assigned a four-character abbreviation based on the short description of the method (fig. 39)—Remove parent and keep children (RPKC), Merge children with parents (MCWP), Remove parent or merge children with parent (RPMC), Distribute parent among children (DPAC), and retain the ambiguities in the original data (ORIG). These four methods are further identified by appending a suffix indicating whether the method is applied to each sample separately ("-S") or for all samples combined ("-C"). Therefore, the method **DPAC-C** distributes parents among children based on ambiguous taxa derived for all samples combined.

## Remove Ambiguous Taxa At or Above

The option to Remove ambiguous taxa at or above is used to remove ambiguous taxa that occur at or above a taxonomic level specified by the user. Ambiguous taxa are identified separately for each sample and removed if they occur at or above the specified taxonomic level. For example, if a sample contains abundance data for Hydropsychidae, *Hydropsyche* sp., and *Hydropsyche betteni* and the user specified family as the criterion for removing ambiguous taxa, the Hydropsychidae data would be removed and data for *Hydropsyche* sp. and *Hydropsyche betteni* would be retained for this sample. This option is executed before the resolve taxonomic ambiguity options.

## Resolve Ambiguities By

The Resolve ambiguities by option is used to resolve ambiguous taxa using the RPKC, MCWP, RPMC, or DPAC methods (fig. 39). These methods can be applied to each sample separately (Keeping samples separate) or for a combined group of samples (Combining all samples). Resolving ambiguities differs from removing ambiguous taxa (Remove ambiguous taxa at or above) by providing more options for identifying ambiguous taxa and combining ambiguous parents with children or children with parents.

## Keeping Samples Separate

The Keeping samples separate option processes each sample separately without considering ambiguities that might exist among samples. It does not deal with situations where a taxon in one sample might be an ambiguous parent or child of taxa in another sample (contrast this approach with the Combining all samples method). The sample-by-sample methods (RPKC-S, MCWP-S, RPMC-S and DPAC-S) are most appropriately used when there is no prior expectation that the sites in the dataset will have similar assemblages in the absence of anthropogenic effects. These methods would be appropriate for analyses across large geographic areas or across areas with complex environmental settings. The five options that are available under the sample-by-sample method have varying effects on how much taxa richness and abundance is preserved (table 36). The choice of option will be based on how comfortable the analyst is with losing richness and (or) abundance in the process of making datasets as comparable as possible.

**Option 1: Remove Parent and Keep Children (RPKC-S).**

Processing method **RPKC-S** identifies the parents that are ambiguous (that is, exist as BU_IDs and as the parent of another entry in the BU_ID column) in each sample and then deletes them. The children of the ambiguous parent are not modified. The results of this method are fewer taxa (lower richness) and lower total abundance in a sample that has one or more ambiguous parents. It is most appropriately used in the analysis of qualitative samples when the user is interested in comparing taxa richness among sites and does not consider

**Table 36.** Taxa richness and abundance obtained by using different methods for resolving taxonomic ambiguities.

[Each method was applied to the four samples listed in table 35. The upper taxa levels in methods MCWP-S and MCWP-C were set to phylum]

| Method | Taxa richness | | | | Abundance | | | |
|--------|------|------|------|------|------|------|------|------|
| | 7650 | 7896 | 7729 | 7973 | 7650 | 7896 | 7729 | 7973 |
| ORIG | 11 | 10 | 11 | 6 | 322 | 312 | 246 | 283 |
| RPKC-S | 8 | 7 | 8 | 5 | 212 | 104 | 231 | 239 |
| MCWP-S | 5 | 2 | 2 | 4 | 322 | 312 | 246 | 283 |
| RPMC-S | 8 | 5 | 8 | 5 | 212 | 304 | 231 | 239 |
| DPAC-S | 8 | 7 | 8 | 5 | 322 | 312 | 246 | 283 |
| RPKC-C | 11 | 7 | 8 | 6 | 212 | 104 | 231 | 239 |
| MCWP-C | 5 | 2 | 2 | 1 | 322 | 312 | 246 | 283 |
| RPMC-C | 6 | 6 | 8 | 4 | 132 | 148 | 231 | 209 |
| DPAC-C | 11 | 7 | 8 | 6 | 322 | 312 | 246 | 283 |

abundances. It is less useful in situations when abundance is important because the application of this method can lead to a substantial loss of abundance.

To show how the **RPKC-S** option works, it was applied to the data in table 35, which includes lifestage information. Ephemeroptera L and Baetidae L were dropped from all samples because children exist in the samples where these taxa occur. Ephemeroptera A was not dropped because all of the children associated with Ephemeroptera are larvae (L) rather than adults (A). Consequently, Ephemeroptera A is not an ambiguous taxon, but Ephemeroptera L is an ambiguous taxon. If the dataset had been processed with the option to combine lifestages, then Ephemeroptera would not appear in the processed data. *Acentrella* sp. L, *Baetis* sp. L, *Plauditus* sp. L, *Pseudocloeon* sp. L, and Zygoptera L were dropped in the samples where children exist, but they were retained in the processed dataset because they occur without children (that is, they are not ambiguous taxa) in several of the samples. The abundance and taxa richness of the processed samples (table 37) are substantially reduced from those observed in the original data (table 35). Taxonomic ambiguities were removed from each sample but not from the dataset. To fully understand how this processing option works, contrast table 37 with the results obtained using the equivalent resolution method under the "combined" option (method **RPKC-C**).

## Option 2: Merge Children with Parents (MCWP-S).

Method **MCWP-S** identifies the ambiguous parents, the children associated with each ambiguous parent, and the sum of the abundances of the children associated with each ambiguous parent. The IDAS program then adds the children's abundances to the appropriate ambiguous parent and deletes the associated children. The program does this iteratively, starting at species and progressing up to phylum. This method will result in fewer taxonomic entities (richness), but total abundance will remain unchanged (table 38). The **MCWP-S** method is appropriate when the user wishes to preserve abundance at the expense of taxa richness. This is an extremely conservative approach that can result in summarizing the data into a relatively small number of fairly high taxonomic levels. Method **MCWP-S** should be used with great caution because it eliminates much of the information content of the dataset.

The presence of just a few taxa with the BU_ID identifier at the order or class level can cause method **MCWP-S** to aggregate data into a few high taxonomic levels that are not very useful for analysis. This problem arises when a dataset contains organisms that have been identified at taxonomic levels that are much higher than the taxonomic levels specified in the sample processing-protocol (Moulton and others, 2000). Typically, this problem occurs when the NWQL BG identifies

**Table 37.** Results obtained by using processing method RPKC-S (removing ambiguous parents and retaining children separately for each sample) to resolve ambiguous taxa in table 35.

[Occurrence is the number of sites where a taxon occurs (1 sample per site). Taxa are listed by name and lifestage (A = adult, P = pupa, L = larvae)]

| Taxon and lifestage | Occurrence | Samples | | | |
|---|---|---|---|---|---|
| | | 7650 | 7896 | 7729 | 7973 |
| Ephemeroptera A | 1 | 2 | | | |
| *Acentrella* sp. L | 1 | | | | 30 |
| *Acentrella parvula* L | 3 | 26 | 11 | 34 | |
| *Acentrella turbida* L | 2 | | 9 | 3 | |
| *Baetis* sp. L | 1 | 35 | | | |
| *Baetis flavistriga* L | 2 | | 6 | 54 | |
| *Baetis intercalaris* L | 2 | | | 78 | 21 |
| *Baetis pluto* L | 3 | | 12 | 23 | 23 |
| *Baetis tricaudatus* L | 2 | | 2 | 11 | |
| *Plauditus* sp. L | 1 | | 56 | | |
| *Plauditus cestus* L | 2 | 34 | | | 53 |
| *Pseudocloeon* sp. L | 1 | 45 | | | |
| *Pseudocloeon propinquum* L | 2 | | | 5 | 112 |
| Zygoptera L | 1 | 10 | | | |
| *Argia* sp. L | 1 | | 8 | | |
| Hydropsychidae L | 1 | 50 | | | |
| *Diplectrona* sp. P | 2 | 10 | | 23 | |
| Total richness | | 8 | 7 | 8 | 5 |
| Total abundance | | 212 | 104 | 231 | 239 |

**Table 38.** Results obtained by using processing method MCWP-S (deleting children of ambiguous parents and adding their abundances to that of the parents) to resolve ambiguous taxa in table 35.

[The upper taxa limit was set to phylum, which conserves the abundance in each sample while reducing the taxa richness. Occurrence is the number of sites where a taxon occurs (one sample per site). Taxa are listed by name and lifestage (A = adult, P = pupa, L = larvae)]

| Taxon and lifestage | Occurrence | Sample | | | |
|---|---|---|---|---|---|
| | | 7650 | 7896 | 7729 | 7973 |
| Ephemeroptera A | 1 | 2 | | | |
| Ephemeroptera L | 1 | 250 | | | |
| Baetidae L | 2 | | 204 | 223 | |
| *Acentrella* sp. L | 1 | | | | 30 |
| *Baetis* sp. L | 1 | | | | 88 |
| *Plauditus cestus* L | 1 | | | | 53 |
| *Pseudocloeon propinquum* L | 1 | | | | 112 |
| Zygoptera L | 2 | 10 | 108 | | |
| Hydropsychidae L | 1 | 50 | | | |
| *Diplectrona* sp. P | 2 | 10 | | 23 | |
| Taxa richness | | 5 | 2 | 2 | 4 |
| Total abundance | | 322 | 312 | 246 | 283 |

fragments of invertebrates in an effort to provide as much information about the contents of a sample as possible. The IDAS program addresses this problem by allowing the user to select an upper taxonomic limit (note: the default for the upper taxa limit is family) above which it will not combine children (fig. 43). A pop-up window allows the user to specify whether to delete or keep ambiguous parents that occur at taxonomic levels above the user-specified upper taxa level (fig. 44). This process prevents method **MCWP-S** from aggregating ambiguous taxa into a few, very high taxonomic levels. Non-ambiguous taxa that occur above the upper taxonomic limit are not affected.

The upper taxa limit selected for aggregation can have a profound effect on the results, both in richness and abundance, obtained by using method **MCWP-S** as illustrated in table 39, which shows how selecting various upper taxa limits affect the processed dataset. For example, setting the upper taxa limit to genus results in the elimination of ambiguous parents that are above this taxonomic level (that is, Arthropoda L, Insecta L, Ephemeroptera L, Leptophlebiidae L, and Hydropsychidae L), but ambiguous parents that are at this taxonomic limit (*Acentrella* sp. and *Plauditus* sp.) are retained. Non-ambiguous taxa that are above the upper taxa limit (Turbellaria) are not eliminated. As the upper taxa limit increases, ambiguous taxa are aggregated into a smaller number of taxonomic entities, and taxa richness decreases. In contrast, the proportion of original abundance that remains in the processed dataset increases as the upper taxa limit increases because fewer taxa are being eliminated and more taxa are being aggregated. When the upper taxa limit is equivalent to the highest taxonomic level associated with an ambiguous parent (phylum in

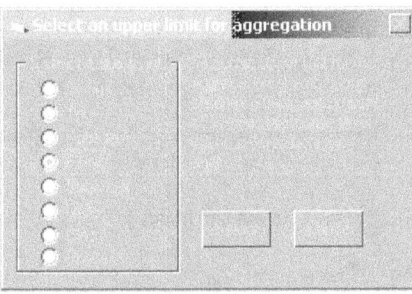

**Figure 43.** Pop-up window that prompts the user to select an upper taxonomic limit for the aggregation of data when using ambiguous taxa resolution methods MCWP-S and MCWP-C.

**Figure 44.** Pop-up window that allows the user to delete or keep ambiguous parents that occur at taxonomic levels above the user-specified upper taxa level.

**Table 39.** Results obtained by specifying different upper taxa limits in the MCWP-S processing method.

[Ambiguous parents are identified (Ambig = Yes) in the original data]

| Taxon and lifestage | Taxonomic level | Original data | | Upper taxa limit | | | | |
|---|---|---|---|---|---|---|---|---|
| | | Value | Ambig | Genus | Family | Order | Class | Phylum |
| Turbellaria | Class | 14 | | 14 | 14 | 14 | 14 | 14 |
| Arthropoda L | Phylum | 10 | Yes | | | | | 389 |
| Insecta L | Class | 25 | Yes | | | | 379 | |
| Ephemeroptera A | Order | 2 | | 2 | 2 | 2 | 2 | 2 |
| Ephemeroptera L | Order | 100 | Yes | | | 318 | | |
| Leptophlebiidae L | Family | 65 | Yes | | 68 | | | |
| *Habrophlebia vibrans* L | Species | 3 | | 3 | | | | |
| *Acentrella* sp. L | Genus | 5 | Yes | 31 | 31 | | | |
| *Acentrella parvula* L | Genus | 26 | | | | | | |
| *Baetis* sp. L | Genus | 35 | | 35 | 35 | | | |
| *Plauditus* sp. L | Genus | 5 | Yes | 39 | 39 | | | |
| *Plauditus cestus* L | Species | 34 | | | | | | |
| *Pseudocloeon* sp. L | Genus | 45 | | 45 | 45 | | | |
| Zygoptera L | SubOrder | 10 | | 1 | 1 | 1 | | |
| Hydropsychidae L | Family | 23 | Yes | | 26 | 26 | | |
| *Diplectrona modesta* L | Species | 2 | | 2 | | | | |
| *Ceratopsyche alhedra* L | Species | 1 | | 1 | | | | |
| Taxa richness | | 17 | | 10 | 9 | 5 | 3 | 3 |
| Total abundance | | 405 | | 173 | 261 | 361 | 395 | 405 |

table 39), then abundance is conserved (that is, the abundance in the processed data is the same as in the original dataset). Users are advised to compare the results obtained by using method **MCWP-S** with the original data so that they fully understand how this method alters their data.

## Option 3: Remove Parent or Merge Children with Parent (RPMC-S).

Method **RPMC-S** identifies ambiguous parents and the children associated with them and compares the abundance of the ambiguous parents to the sum of the abundances of their children. If the abundance of the ambiguous parent is greater than the sum of the children's abundances, the children's abundances are added to the parent abundance and the children are deleted. Otherwise, the parent is deleted and the children are retained. Ambiguities are resolved iteratively, starting with genus and progressing up to phylum. The IDAS program keeps track of abundances that are deleted at each taxonomic level and can add these abundances to the parent in later iterations. This process would be necessary if a comparison between an ambiguous parent and its children indicates that the children need to be combined with the parent and one or more parents have been deleted in a previous iteration. Adding the abundances of deleted parents under these circumstances prevents the loss of abundance information. This method results in both reduced sample richness and abundance.

Method **RPMC-S** offers a compromise between methods **RPKC-S**, which eliminates ambiguous parents, and **MCWP-S**, which eliminates the children of ambiguous parents. This option is used when the analyst wants to tie the preservation of taxa richness to the abundance of the ambiguous taxa and is unwilling to accept the assumptions that are associated with method **DPAC-S**. Method **DPAC-S**, which distributes the abundance of ambiguous parents among the children in accordance with the relative abundance of the children, assumes that the ambiguous parents are composed solely of the ambiguous children that occur in the sample (that is, they cannot constitute an unidentified taxon) and occur in proportion to the relative abundance of the ambiguous children. If the user is uncomfortable with these assumptions, method **RPMC-S** should be used.

Taxonomic ambiguities are resolved iteratively in method **RPMC-S**, starting with species and working up to family in the example presented in table 40. The lowest taxonomic level (BU_ID column) is genus, so the first iteration resolves ambiguities between tribe and genus. Both (Chironomini and Tanytarsini) tribes are ambiguous; that is, each has three genera associated with it. The combined abundance of the children of Chironomini (*Chernoushii* (10), *Chironomus* (20), and *Cladopelma* (5)) is less than the abundance of Chironomini (100) so the abundances of the children are combined with the parent (10+20+5+100=135) and the child abundances

**Table 40.**    An example of how processing method RPMC-S resolves ambiguous taxa over three iterations from genus to family.

| Family | Subfamily | Tribe | Genus (BU_ID) | Original abundance | Iteration 1: Tribe | Iteration 2: Subfamily | Iteration 3: Family |
|---|---|---|---|---|---|---|---|
| Chironomidae | | | | 100 | 100 | 100 | 0 |
| | Chironominae | | | 225 | 225 | 430 | 430 |
| | | Chironomini | | 100 | 135 | 0 | 0 |
| | | | *Chernoushii* | 10 | 0 | 0 | 0 |
| | | | *Chironomus* | 20 | 0 | 0 | 0 |
| | | | *Cladopelma* | 5 | 0 | 0 | 0 |
| | | Tanytarsini | | 5 | 0 | 0 | 0 |
| | | | *Cladotanytarsus* | 15 | 15 | 0 | 0 |
| | | | *Microspectra* | 20 | 20 | 0 | 0 |
| | | | *Neozaurelia* | 30 | 30 | 0 | 0 |
| | Diamesinae | | | 10 | 10 | 0 | 0 |
| | | | *Diamesa* | 15 | 15 | 15 | 15 |
| | | | *Pagastia* | 35 | 35 | 35 | 35 |
| | | | *Potthastia* | 50 | 50 | 50 | 50 |
| | | | Total resolved | 640 | 635 | 630 | 530 |
| | | | Carryover | 0 | 5 | 10 | 110 |
| | | | Total processed | 640 | 640 | 640 | 640 |

are set to zero. The opposite is true for the Tanytarsini; the sum of the abundances of the children (65) is greater than the abundance of the ambiguous parent (5). Therefore, the parent's abundance is set to zero and the children's abundances are retained. The abundance of the Tanytarsini is transferred to a "carry over" variable so that this abundance is not lost if the children are combined with a parent at a higher taxonomic level in a later iteration. These results are summarized in the "Iteration 1" column of table 40.

The second iteration resolves ambiguities between the first iteration (tribe) and the next taxonomic level (subfamily). Two ambiguous subfamilies are present in this dataset—Chironominae and Diamesinae. The abundance of Chironominae (225) is greater than the abundance of the children (Chironomini + *Cladotanytarsus* + *Microspectra* + *Neozaurelia* = 200) so the children's abundances are added to the parents (225+200=425). However, the carryover of Tanytarsini (5) from the previous iteration causes the abundance of Chironominae to become 425+5=430, and the carryover is set to zero. In the case of Diamesinae, the abundance of the children (100) is greater than the ambiguous parent's abundance, so the abundance of Diamesinae is set to zero and the abundance of Diamesinae (10) is carried over to the next iteration.

Iteration 3 resolves ambiguities between the subfamily and family levels. In this example the abundance of Chironomidae (100) is less than the sum of the abundance of children (530), so the abundance of Chironomidae is set to zero and the carryover value becomes 110 (100 Chironomidae + 10 Diamesinae). Resolving ambiguities by using option 3 reduces both the number of taxa and the total abundance in the sample. Results of applying method **RPMC-S** to the four hypothetical samples presented in table 35 are shown in table 41. These results clearly show that method RPMC-S is a compromise between methods **RPKC-S** and **MCWP-S**. Method **RPMC-S** preserves more of the original taxa richness than does method **MCWP-S** (table 38), but less than method **RPKC-S** (table 37). In contrast, method **RPMC-S** preserves more of sample abundance than does method **RPKC-S** but less than method **MCWP-S**. As with all methods for resolving ambiguities, the user is advised to compare the results with the original data so that the user fully understands how this method has modified the original data before calculating community metrics or exporting data for analysis.

**Table 41.** Results obtained by using processing method RPMC-S to resolve ambiguous taxa in table 35.

[Occurrence is the number of sites where a taxon occurs (one sample per site). Taxa are listed by name and life-stage (A = adult, P = pupa, L = larvae)]

| Taxon and lifestage | Occurrence | Sample | | | |
|---|---|---|---|---|---|
| | | 7650 | 7896 | 7729 | 7973 |
| Ephemeroptera A | 1 | 2 | | | |
| *Acentrella* sp. L | 1 | | | | 30 |
| *Acentrella parvula* L | 3 | 26 | 11 | 34 | |
| *Acentrella turbida* L | 2 | | 9 | 3 | |
| *Baetis* sp. L | 2 | 35 | 120 | | |
| *Baetis flavistriga* L | 1 | | | 54 | |
| *Baetis intercalaris* L | 2 | | | 78 | 21 |
| *Baetis pluto* L | 2 | | | 23 | 23 |
| *Baetis tricaudatus* L | 1 | | | 11 | |
| *Plauditus* sp. L | 1 | | 56 | | |
| *Plauditus cestus* L | 2 | 34 | | | 53 |
| *Pseudocloeon* sp. L | 1 | 45 | | | |
| *Pseudocloeon propinquum* L | 2 | | | 5 | 112 |
| Zygoptera L | 2 | 10 | 108 | | |
| Hydropsychidae L | 1 | 50 | | | |
| *Diplectrona* sp. P | 2 | 10 | | 23 | |
| Taxa richness | | 8 | 5 | 8 | 5 |
| Total abundance | | 212 | 304 | 231 | 239 |

## Option 4: Distribute Parent Among Children (DPAC-S).

Method **DPAC-S** identifies ambiguous parents, the children associated with each ambiguous parent, and the sum of the abundances of the children associated with each ambiguous parent. It then distributes the abundances of the ambiguous parents among the children in accordance with the relative abundance of each child ($C_i / \Sigma\ C_i$, where $C_i$ is the abundance of the ith child). The ambiguous parent is then deleted. Ambiguities are resolved iteratively, starting at species and moving up to phylum as shown in table 42 for sample 7896 (from table 35). The first iteration distributes the abundance of *Baetis* sp. L among the three species of *Baetis* (*flavistriga, pluto,* and *tricaudatus*) in accordance with the relative abundance of each child. For example, the relative abundance of *Baetis flavistriga* L is 0.3 (6/[6+12+2]) so 30 percent of the abundance of *Baetis* sp. L (100 * 0.3 = 30) is added to the abundance of *Baetis flavistriga* (30+6 = 36). The updated abundances are then used as the input for iteration 2, which distributes the abundance associated with the ambiguous family Baetidae (8) among the six ambiguous children. As before, this abundance is distributed in accordance with the relative abundance of the children

(for example, the relative abundance of *Baetis flavistriga* L is 36/[11+9+36+72+12+56] = 0.184). The abundance of *Baetis flavistriga* L after iteration 2 is 37.5 (36+[8*0.184]). Iteration 3 distributes the abundance of Zygoptera L (100) among the children associated with this ambiguous parent. Because there is only one child associated with Zygoptera L (*Argia* sp. L), all of the abundance is added to *Argia* sp. L (8+100=108). Processing data by using this method results in reduced rows (richness) in the processed dataset, but the total abundance in each sample remains the same (table 43).

This method is used when the analyst wants to preserve as much taxa richness and abundance as possible provided that the analyst is comfortable with the assumptions used in this method to resolve ambiguities. This method assumes that the ambiguous parents are composed only of the taxa that make up the ambiguous children and that the abundance of the ambiguous parent is composed of the children in the same proportions as they occur in the sample. If the user is unwilling to accept these assumptions and wants to preserve as much taxa richness and abundance as possible, then method **RPMC-S** should be used rather than method **DPAC-S**.

**Table 42.**    Examples of how processing method DPAC-S distributes the abundances of ambiguous parents among ambiguous children for sample 7896 in table 35.

| Iteration 1: Distribute abundance of *Baetis* sp. L (100) among children | | | | |
|---|---|---|---|---|
| Taxon | Original abundance | Percentage of child's abundance | Abundance to be distributed | Abundances after iteration 1 |
| Baetidae L | 8 | | | 8 |
| *Acentrella parvula* L. | 11 | | | 11 |
| *Acentrella turbida* L | 9 | | | 9 |
| *Baetis* sp. L | 100 | | | 0 |
| *Baetis flavistriga* L | 6 | 30.0 | 30.0 | 36 |
| *Baetis pluto* L | 12 | 60.0 | 60.0 | 72 |
| *Baetis tricaudatus* L | 2 | 10.0 | 10.0 | 12 |
| *Plauditus* sp. L | 56 | | | 56 |
| Zygoptera L | 100 | | | 100 |
| *Argia* sp. L | 8 | | | 8 |

| Iteration 2: Distribute abundance of Baetidae L (8) among children | | | | |
|---|---|---|---|---|
| Taxon | Abundances after iteration 1 | Percentage of child's abundance | Abundance to be distributed | Abundances after iteration 2 |
| Baetidae L | 8 | | | 0.0 |
| *Acentrella parvula* L. | 11 | 5.6 | 0.4 | 11.4 |
| *Acentrella turbida* L | 9 | 4.6 | 0.4 | 9.4 |
| *Baetis* sp. L | 0 | | | 0.0 |
| *Baetis flavistriga* L | 36 | 18.4 | 1.5 | 37.5 |
| *Baetis pluto* L | 72 | 36.7 | 2.9 | 74.9 |
| *Baetis tricaudatus* L | 12 | 6.1 | 0.5 | 12.5 |
| *Plauditus* sp. L | 56 | 28.6 | 2.3 | 58.3 |
| Zygoptera L | 100 | | | 100.0 |
| *Argia* sp. L | 8 | | | 8.0 |

| Iteration 3: Distribute abundance of Zygoptera L (100) among children | | | | |
|---|---|---|---|---|
| Taxon | Abundances after iteration 2 | Percentage of child's abundance | Abundance to be distributed | Abundances after iteration 3 |
| Baetidae L | 0.0 | | | 0.0 |
| *Acentrella parvula* L. | 11.4 | | | 11.4 |
| *Acentrella turbida* L | 9.4 | | | 9.4 |
| *Baetis* sp. L | 0.0 | | | 0.0 |
| *Baetis flavistriga* L | 37.5 | | | 37.5 |
| *Baetis pluto* L | 74.9 | | | 74.9 |
| *Baetis tricaudatus* L | 12.5 | | | 12.5 |
| *Plauditus* sp. L | 58.3 | | | 58.3 |
| Zygoptera L | 100.0 | | | 0.0 |
| *Argia* sp. L | 8.0 | 100.0 | 100 | 108.0 |

**Table 43.** Results obtained by using processing method DPAC-S (distributing the abundance of ambiguous parents among their children in accordance with the relative abundance of each child) to resolve ambiguous taxa in table 35.

[Occurrence is the number of sites where a taxon occurs (one sample per site). Taxa are listed by name and lifestage (A = adult, P = pupa, L = larvae)]

| Taxon and lifestage | Occurrence | Sample | | | |
|---|---|---|---|---|---|
| | | 7650 | 7896 | 7729 | 7973 |
| Ephemeroptera A | 1 | 2.0 | | | |
| *Acentrella* sp. L | 1 | | | | 30.0 |
| *Acentrella parvula* L | 3 | 51.7 | 11.4 | 44.0 | |
| *Acentrella turbida* L | 2 | | 9.4 | 3.9 | |
| *Baetis* sp. L | 1 | 58.3 | | | |
| *Baetis flavistriga* L | 2 | | 37.5 | 55.0 | |
| *Baetis intercalaris* L | 2 | | | 79.4 | 42.0 |
| *Baetis pluto* L | 3 | | 74.9 | 23.4 | 46.0 |
| *Baetis tricaudatus* L | 2 | | 12.5 | 11.2 | |
| *Plauditus* sp. L | 1 | | 58.3 | | |
| *Plauditus cestus* L | 2 | 65.0 | | | 53.0 |
| *Pseudocloeon* sp. L | 1 | 75.0 | | | |
| *Pseudocloeon propinquum* L | 2 | | | 6.1 | 112.0 |
| Zygoptera L | 1 | 10.0 | | | |
| *Argia* sp. L | 1 | | 108.0 | | |
| Hydropsychidae L | 1 | 50.0 | | | |
| *Diplectrona* sp. P | 2 | 10.0 | | 23.0 | |
| Taxa richness | | 8 | 7 | 8 | 5 |
| Total abundance | | 322.0 | 312.0 | 246.0 | 283.0 |

**Option 5: None – Retain Ambiguities (ORIG).**

The user has the option of not resolving taxonomic ambiguities. This option is useful when the analyst wants to compare the original data with the processed data to fully understand how the data-preparation options have modified the data. Data processed by using method **ORIG** can be processed by the other IDAS modules so the effects of data-preparation choices on assemblage metrics, diversities, similarities, or other analyses can be ascertained quickly. The tab-delimited, full format option of the Data Export module provides an efficient mechanism for viewing and comparing datasets. When the **ORIG** method is selected, the IDAS program will apply all of the options selected by the user and will generate a new Excel spreadsheet or Access data table in the processed format (table 1). The transformation to the processed format includes the deletion of all data in the **Notes** column (BG processing notes), because combining data on the basis of BU_ID or BU_ID+lifestage may break the association between **Notes** and specific lines of data.

## Combining All Samples

The IDAS program allows the user to resolve taxonomic ambiguities for a block of samples rather than for each sample separately (Combining all samples). This produces a dataset in which there are no ambiguous taxa within the samples or among samples. In contrast, the sample-by-sample methods (**RPKC-S, MCWP-S, RPMC-S** and **DPAC-S**) resolve taxonomic ambiguities in each sample, but ambiguities may still exist among samples within the dataset (tables 37, 38,

> **TIP: The "Combined" method of resolving ambiguous taxa should only be used when there is an expectation that all sites will have similar communities (that is, for geographically small areas that are environmentally similar). This option is not appropriate for comparing groups of sites across broad geographic areas where communities would be expected to differ markedly even in the absence of human effects.**

41, and 43). The combined methods are advantageous when analyzing datasets in which the assemblages are expected to be similar in the absence of human effects. This approach is not appropriate for situations where communities are expected to have large differences that are related to natural factors.

The combined methods (**RPKC-C, MCWP-C, RPMC-C** and **DPAC-C**) create a new sample that is a combination of all the other samples. Ambiguous parents and the children associated with them are identified in the combined sample (table 35). Information derived from the combined sample is then used to resolve taxonomic ambiguities in each sample. The actual methods for resolving ambiguities are similar to those used for resolving ambiguities in separate samples. However, the ambiguous taxa identified by using the combined methods can be quite different from the ambiguities that are identified for individual samples (table 35), and the taxa richness and abundances extracted from the samples can be quite different from those obtained by using the "separate" methods (table 36; fig. 42).

The combined methods are most useful in situations where the user expects a group of sites to have the same assemblages in the absence of anthropogenic effects. The identities of missing children can then be estimated based on the children present in the combined datasets. Urban gradient studies, which rely on the selection of sites with similar natural environmental characteristics, are examples of such a situation. Resolving ambiguities by combining samples reduces differences among sites as compared to resolving ambiguities on a sample-by-sample basis where immature or damaged specimens in a sample may lead to discrepancies among samples. This approach is not suitable for analyzing data from a broad geographic area, multiple Study Units, or any instance where the communities are expected to differ substantially among sites because of natural factors.

**Option 1: Remove Parent and Keep Children (RPKC-C).**

Method **RPKC-C** combines all of the samples together and then identifies parents that are ambiguous (that is, exist as BU_IDs and as the parent of another entry in the BU_ID column of the combined dataset) based on the combined data (see **Combined samples** column in table 35). Ambiguous parents are then deleted in each sample on the basis of whether they were identified as ambiguous parents in the combined dataset rather than in the individual samples. There can be substantial differences in the taxa identified as ambiguous depending on whether the determination was made by using combined data or individual samples (table 35). One consequence of resolving ambiguities based on the combined data is that some samples may contain ambiguous parents but no associated children (for example, *Baetis* sp. L in sample 7650, table 35). The IDAS program addresses this problem by calling up a subroutine that gives the user three options for resolving these situations: (1) associate one or more children with the ambiguous parent, (2) delete the ambiguous parent, or (3) retain the

> **TIP:** Do not resolve ambiguous taxa by using the "combined samples" option if the dataset contains a mixture of qualitative (QMH or QUAL) and quantitative (RTH or DTH) samples. The results can be misleading because the qualitative data always have a value of one. Qualitative and quantitative samples should be processed separately.

ambiguous parent (see **Associating ambiguous children with ambiguous parents**). These choices are applied to all samples where the ambiguous parent is missing. They are not applied to situations where the sample has one or more children that are associated with the ambiguous parent. If the dataset has only one child that can be associated with the ambiguous parent (for example, *Plauditus cestus* L, sample 7650, table 35), then the IDAS program automatically associates that child with the ambiguous parent without informing the user. Information on the assignment of children to parents is stored in an Excel worksheet or Access data table that ends in the suffix "_PC_Assoc" (table 44, ALBE_PC_Assoc).

Resolving ambiguities by using method **RPKC-C** results in fewer taxonomic entities (richness) and lower total abundance in samples that have one or more ambiguous parents (table 45). This option is most useful in the analysis of qualitative samples, where the user is only interested in comparing taxa richness among sites. It is less useful in situations where comparisons of abundance are important, because the application of this method can lead to a substantial loss of abundance. The advantage of using the combined methods becomes apparent upon comparison of the taxa lists generated by method **RPKC-S** (table 37) and **RPKC-C** (table 45) with the original data (table 35). Each sample processed by using method **RPKC-S** is free of ambiguous taxa; however, the combined taxa list includes taxa that are ambiguous (for example, both *Acentrella* sp. L and *Acentrella parvula* L are present), whereas the list generated by **RPKC-C** has no ambiguities. This is advantageous when comparing data from sites where the underlying assemblages are expected to contain similar taxa in the absence of anthropogenic effects. However, this advantage must be balanced against situations where the user is required to assign children to ambiguous parents, which constitutes an estimation of missing data. The user is advised to consider how much of the data are estimated before using this technique.

**Table 44.**   The division of ambiguous parents among children in methods RPKC-C and DPAC-C is documented in a spreadsheet or table with the suffix "_PC_Assoc".

[Example 1: Only one child, *Ferrissia* sp. exists for Ancylidae so all the parent's abundance is assigned to that child. Example 2: Two children, *Stenelmis* sp. and *Psephenus* sp. are associated with the ambiguous parent Coleoptera. The abundance of the Coleoptera is divided among the two children in accordance with their relative abundance (0.86 and 0.14, respectively)]

| Column | Example 1 | Example 2a | Example 2b | Description |
|---|---|---|---|---|
| pTkey | Ancylidae | Coleoptera | Coleoptera | Parent identifier (key) |
| pBU_ID | Ancylidae | Coleoptera | Coleoptera | parent BU_ID |
| pLifestage | | | | parent lifestage |
| pSortCode | 8000125 | 8002419 | 8002419 | Parent sort code |
| cTkey | *Ferrissia* sp. | *Stenelmis* sp. | *Psephenus* sp. | Child identifier (key) |
| cBU_ID | *Ferrissia* sp. | *Stenelmis* sp. | *Psephenus* sp. | Child BU_ID |
| cLifestage | | L | L | Child lifestage |
| cSortCode | 8000126 | 8002810 | 8002846 | Child sort code |
| cPercent | 1 | 0.86 | 0.14 | Child's percent abundance |

**Table 45.**   Results obtained by using processing method RPKC-C (remove ambiguous parents and keep children) to resolve ambiguous taxa in table 35.

[Parent abundance is distributed across all children of the parent. Occurrence is the number of sites where a taxon occurs (one sample per site). Taxa are listed by name and lifestage (A = adult, P = pupa, L = larvae)]

| Taxon and lifestage | Occurrence | Sample | | | |
|---|---|---|---|---|---|
| | | 7650 | 7896 | 7729 | 7973 |
| Ephemeroptera A | 1 | 2.0 | | | |
| *Acentrella parvula* L | 4 | 26.0 | 11.0 | 34.0 | 25.7 |
| *Acentrella turbida* L | 2 | | 9.0 | 3.0 | 4.3 |
| *Baetis flavistriga* L | 3 | 9.1 | 6.0 | 54.0 | |
| *Baetis intercalaris* L | 2 | 15.1 | | 78.0 | 21.0 |
| *Baetis pluto* L | 3 | 8.8 | 12.0 | 23.0 | 23.0 |
| *Baetis tricaudatus* L | 3 | 2.0 | 2.0 | 11.0 | |
| *Plauditus cestus* L | 3 | 34.0 | 56.0 | | 53.0 |
| *Pseudocloeon propinquum* L | 3 | 45.0 | | 5.0 | 112.0 |
| *Argia* sp. L | 2 | 10.0 | 8.0 | | |
| Hydropsychidae L | 1 | 50.0 | | | |
| *Diplectrona* sp. P | 2 | 10.0 | | 23.0 | |
| Taxa richness | | 11.0 | 7.0 | 8.0 | 6.0 |
| Total abundance | | 212.0 | 104.0 | 231.0 | 239.0 |

**Option 2: Merge Children With Parents (MCWP-C).**

Method **MCWP-C** is similar to method **MCWP-S** except that ambiguous parents are identified on the basis of the combined data (table 35) rather than individually for each sample. Once the ambiguous parents have been identified, the IDAS program determines which children are associated with each ambiguous parent and calculates the sum of the children's abundances. The program then combines the children's abundances with the parent's abundances iteratively, starting at species and progressing up to phylum. As with method **MCWP-S**, the presence of damaged specimens can result in aggregating abundances into just a few, very high taxonomic levels (for example, class or order), which may not be desirable for subsequent analyses. To avoid this problem, the IDAS program allows the user to specify an upper taxonomic limit (fig. 43) above which it will not combine children (see

MCWP-S). This is accomplished by identifying ambiguous parents that occur at taxonomic levels higher than the upper taxonomic limit specified by the user (note: the default value for the upper taxonomic limit is family). A pop-up window allows the user to specify whether to delete or keep ambiguous parents that occur at taxonomic levels above the user-specified upper taxa level (fig. 44). The selection of an upper taxonomic limit can greatly affect taxa richness and abundance (see example for **MCWP-S**, table 39) so the user should understand how the level chosen affects the data generated by this method. Once the IDAS program has identified the ambiguous children and parents for the combined data, it will apply this information to each sample individually. That is, it will add the abundance of the ambiguous children to the appropriate ambiguous parent and then delete the children (table 46). If the appropriate ambiguous parent is not present in the sample, the ambiguous parent will be added to the taxa that constitute the sample, and the abundances of the ambiguous children will be added to it before they are deleted. This method will result in fewer taxonomic entities (richness) but total abundance will remain unchanged, unless the selection of upper taxonomic limit leads to the deletion of data.

Method **MCWP-C** is an extremely conservative method of resolving taxonomic ambiguities that usually results in summarizing taxa richness into a relatively small number of fairly high taxonomic levels. This option is appropriate for analyses when the user places a premium on preserving abundance at the expense of taxa richness and wants to eliminate ambiguities from the entire dataset. It is advised that this option be used with great caution because it can eliminate much of the information contained in a dataset.

## Option 3: Remove Parent or Merge Children with Parent (RPMC).

Method **RPMC-C** is analogous to method **RPMC-S** and produces results that are intermediate between methods **RPKC-C** and **MCWP-C**. This approach combines all the samples into one, identifies the ambiguous parents in the combined sample, calculates the sum of the abundances of the children associated with each ambiguous parent, and then determines whether to delete the parent or children by comparing the sum of the abundances of the associated children with the abundance of the parent. These comparisons are made iteratively, starting with species and processing through phylum. Parents are deleted if the sum of the children's abundances is greater than or equal to the abundance of the parent. Otherwise, the children's abundances are added to the parent and the children are deleted. The IDAS program keeps track of abundances that are deleted so that they can be added back in if the decision is made to add children to the parents at a higher taxonomic level (that is, the deleted parent is a child of an ambiguous parent at the higher taxonomic level). The decisions on which taxa to combine and which to delete are based on the combined sample and then applied to each sample individually.

Applying these decisions across the samples can cause samples to lose or gain taxonomic information depending on whether ambiguous parents or children are present in the sample (table 47). For example, if the decision is made to add children's abundances to an ambiguous parent and the ambiguous parent does not exist in a sample (*Ceratopsyche* sp. L in sample 1, table 47), the parent's name is added to the sample along with the sum of the children's abundances. Conversely,

**Table 46.** Results obtained by using processing option MCWP-C (merge the abundance of children with the abundance of the ambiguous parents) to resolve ambiguous taxa in table 35.

[The upper taxa limit was set to phylum, which conserves the abundance in each sample while reducing the taxa richness. Occurrence is the number of sites where a taxon occurs (one sample per site). Taxa are listed by name and lifestage (A = adult, P = pupa, L = larvae)]

| Taxon and lifestage | Occurrence | Sample | | | |
|---|---|---|---|---|---|
| | | 7650 | 7896 | 7729 | 7973 |
| Ephemeroptera A | 1 | 2 | | | |
| Ephemeroptera L | 4 | 250 | 204 | 223 | 283 |
| Zygoptera L | 2 | 10 | 108 | | |
| Hydropsychidae L | 1 | 50 | | | |
| *Diplectrona* sp. P | 2 | 10 | | 23 | |
| Taxa richness | | 5 | 2 | 2 | 1 |
| Total abundance | | 322 | 312 | 246 | 283 |

**Table 47.** Examples of how processing method RPMC-C resolves ambiguous taxa when parents or children are not present in the sample.

| Taxon and lifestage | Combined samples | | Original data | | Resolved data | |
|---|---|---|---|---|---|---|
| | Total | Ambig | Sample 1 | Sample 2 | Sample 1 | Sample 2 |
| *Baetis* sp. L | 10 | Yes | 8 | 2 | 0 | 0 |
| *Baetis flavistriga* L | 30 | | 0 | 30 | 0 | 30 |
| *Baetis pluto* L | 5 | | 0 | 5 | 0 | 5 |
| *Ceratopsyche* sp. L | 100 | Yes | 0 | 100 | 5 | 102 |
| *Ceratopsyche alhedra* L | 5 | | 4 | 1 | 0 | 0 |
| *Ceratopsyche bronta* L | 2 | | 1 | 1 | 0 | 0 |

if the decision is to delete an ambiguous parent and keep the children and a sample contains the ambiguous parent but no children (*Baetis* sp. L in sample 1, table 47), then the ambiguous parent will be deleted and the abundance of the parent will be lost. The user needs to keep this behavior in mind when using this method because potentially it can lead to the loss of whole groups of organisms.

The application of method **RPMC-C** can have a profound effect on the taxa richness and abundance in the processed sample (table 48) compared to the original sample (table 35) and other methods of resolving taxonomic ambiguities. Taxa richness after applying method **RPMC-C** is less than the original data (**ORIG**), greater than method **MCWP-C**, and less than method **RPKC-C**. Total abundance typically is substantially less than in the original data. Comparison of results obtained by using methods **RPMC-C** (table 48) and **RPMC-S** (table 41) illustrates that the combined methods (**RPKC-C, MCWP-C, RPMC-C, DPAC-C**) for resolving taxonomic ambiguities produce datasets in which there are no taxonomic ambiguities among the samples.

In contrast, the separate methods (**RPKC-S, MCWP-S, RPMC-S, DPAC-S**) produce datasets that have no ambiguities in samples but retain ambiguities among samples. Using the combined methods may be advantageous for analyzing datasets in which the communities are expected to be similar in the absence of anthropogenic effects. The **RPMC-C** method provides a more moderate alternative to methods **RPKC-C** and **MCWP-S** by using the abundance of ambiguous parents and children to determine whether to delete parents or children.

**Option 4: Distribute Parent Among Children (DPAC-C).**

Method **DPAC-C** is similar to method **DPAC-S** in that it resolves an ambiguous parent by distributing its abundance among its children in proportion to the relative abundance of each child. The difference between methods **DPAC-S** and **DPAC-C** is that **DPAC-C** identifies ambiguous taxa based on all samples combined rather than identifying ambiguous taxa for each sample separately (table 35). Once ambiguous

**Table 48.** Results obtained by using processing method RPMC-C to resolve ambiguous taxa in table 35.

[Occurrence is the number of sites where a taxon occurs (one sample per site). Taxa are listed by name and lifestage (A = adult, P = pupa, L = larvae)]

| Taxon and lifestage | Occurrence | Sample | | | |
|---|---|---|---|---|---|
| | | 7650 | 7896 | 7729 | 7973 |
| Ephemeroptera A | 1 | 2 | | | |
| *Acentrella parvula* L | 3 | 26 | 11 | 34 | |
| *Acentrella turbida* L | 2 | | 9 | 3 | |
| *Baetis flavistriga* L | 2 | | 6 | 54 | |
| *Baetis intercalaris* L | 2 | | | 78 | 21 |
| *Baetis pluto* L | 3 | | 12 | 23 | 23 |
| *Baetis tricaudatus* L | 2 | | 2 | 11 | |
| *Plauditus cestus* L | 2 | 34 | | | 53 |
| *Pseudocloeon propinquum* L | 2 | | | 5 | 112 |
| Zygoptera L | 2 | 10 | 108 | | |
| Hydropsychidae L | 1 | 50 | | | |
| *Diplectrona* sp. P | 2 | 10 | | 23 | |
| Taxa richness | | 6 | 6 | 8 | 4 |
| Total abundance | | 132 | 148 | 231 | 209 |

parents and their associated children have been identified for the combined data, this information is used to identify ambiguous parents in each sample separately. The abundance of the ambiguous parent is distributed among the children in accordance with the relative abundance of each child $(C_i / \Sigma C_i$, where $C_i$ is the abundance of the $i^{th}$ child) in the sample iteratively, starting at the species level and processing up to phylum. Once the abundance of an ambiguous parent has been distributed, it is deleted from the sample. Identifying ambiguous taxa based on the combined data can produce a situation in which a sample contains an ambiguous parent but no children (that is, $\Sigma C_i = 0$). This makes it impossible to distribute the abundance of the parent. The IDAS program addresses this problem by calling up a subroutine that gives the analyst three options for handling cases where the ambiguous parent exists but not the children: (1) select children to associate with the ambiguous parent, (2) delete the ambiguous parent, or (3) retain the ambiguous parent (see **Associating Ambiguous Children with Ambiguous Parents**). These choices are applied to all samples in which the ambiguous children are missing. The IDAS program automatically assigns a child to an ambiguous parent if there is only one child associated with the ambiguous parent. The program records which children are assigned to each ambiguous parent and the proportion of the parent's abundance that is assigned to each child. This information is stored in an Excel spreadsheet or Access data table that ends in the suffix "_PC_Assoc."

Applying processing method **DPAC-C** to a dataset results in reduced rows (richness) in the processed dataset compared with unprocessed data (table 35), but the total abundance remains the same (table 49). This method is recommended when the user wants to preserve as much taxa richness and

abundance as possible, provided that the user is comfortable with the assumptions that this method uses to resolve ambiguities. This method assumes that the ambiguous parents are composed only of the associated ambiguous children and that the proportion on the parent's abundance that is composed of each child is the same as the proportion of the children's abundance in the combined sample. The user of this method should also be comfortable with the process of assigning children to ambiguous parents when children are missing in the samples. This constitutes estimating missing data, and analysts need to consider how much of the processed datasets are composed of estimated data when considering which method to use to resolve ambiguous taxa.

## Associating Ambiguous Children with Ambiguous Parents

Methods that resolve ambiguities by combining all samples (**RPKC-C, MCWP-C, RPMC-C, DPAC-C**) identify ambiguous parents and children based on the combined dataset and then apply this information to each sample. Methods that retain children in place of ambiguous parents (**RPKC-C**) or distribute parent abundances among children (**DPAC-C**) can lead to instances where an ambiguous parent exists in a sample but not any of the ambiguous children associated with the ambiguous parent exist (that is, the ambiguous children exist in other samples in the dataset but not in the sample under consideration). Consequently, the sample does not have the information that is needed to transfer the abundance of the ambiguous parent to the ambiguous children. This situation is illustrated in table 35 in which samples 7650, 7896, and 7973 have parents identified as being ambiguous in the combined data (that is, *Baetis* sp. L, *Pseudocloeon* sp. L, and

**Table 49.**   Results obtained by using processing method DPAC-C (distributing the abundance of ambiguous parents among their children) to resolve ambiguous taxa in table 35.

[Occurrence is the number of sites where a taxon occurs (one sample per site). Taxa are listed by name and life-stage (A = adult, P = pupa, L = larvae)]

| Taxon and lifestage | Occurrence | Sample | | | |
|---|---|---|---|---|---|
| | | 7650 | 7896 | 7729 | 7973 |
| Ephemeroptera A | 1 | 2.0 | | | |
| *Acentrella parvula* L | 4 | 51.7 | 11.4 | 44.0 | 30.0 |
| *Acentrella turbida* L | 2 | | 9.4 | 3.9 | |
| *Baetis flavistriga* L | 2 | | 37.5 | 55.0 | |
| *Baetis intercalaris* L | 3 | 36.8 | | 79.4 | 42.0 |
| *Baetis pluto* L | 4 | 21.5 | 74.9 | 23.4 | 46.0 |
| *Baetis tricaudatus* L | 2 | | 12.5 | 11.2 | |
| *Plauditus cestus* L | 3 | 65.0 | 58.3 | | 53.0 |
| *Pseudocloeon propinquum* L | 3 | 75.0 | | 6.1 | 112.0 |
| *Argia* sp. L | 2 | 10.0 | 108.0 | | |
| Hydropsychidae L | 1 | 50.0 | | | |
| *Diplectrona* sp. L | 2 | 10.0 | | 23.0 | |
| Taxa richness | | 9 | 7 | 8 | 5 |
| Total abundance | | 322.0 | 312.0 | 246.0 | 283.0 |

Zygoptera L in sample 7650; *Plauditus* sp. L in sample 7896, and *Acentrell*a sp. L in sample 7973), but the corresponding children are missing from these samples.

The IDAS program keeps track of instances where a sample contains an ambiguous parent but none of the children associated with that parent. In these cases, the user has three options for handling the abundance of the ambiguous parent: (1) retain the ambiguous parent and its abundance, (2) delete the ambiguous parent and its abundance, or (3) select one or more children among which to distribute the parent's abundance. If the ambiguous parent is only associated with a single child in the combined dataset (for example, *Plauditus* sp. L, *Pseudocloeon* sp. L, and Zygoptera L in table 35), the IDAS program automatically assigns the parent's abundance to the child without user intervention. Otherwise, the IDAS program will inform the user of the number of taxa that require user intervention to resolve ambiguities (fig. 45) and give the user the option of processing these taxa or quitting the program.

If the user decides to manually resolve ambiguous taxa that have no children, a new window will open that allows the user to select children to match with the ambiguous parent, retain the ambiguous parent, or delete the ambiguous parent (fig. 46). This window has three data grids. The upper grid (Ambiguous parent) contains the name of the ambiguous parent, lifestage, and statistics on occurrence (percentage of sites and samples where the taxon occurs) and abundance or density (as a percentage of the total abundance or density in the combined data). In the example shown in figure 46, Gastropoda accounts for only a very small percentage of density in the combined samples (0.06) and is present at only a relatively small percentage of sites (15.2) and samples (9.9). Therefore, the user could legitimately decide to delete Gastropoda from consideration (click on Delete parent) because it represents such a small percentage of the combined density and occurs infrequently. If the user chooses to delete Gastropoda, then it will be deleted in all samples in which it is an ambiguous parent without children but not in samples in which children are present. In such cases, the density of Gastropoda will be divided among the associated ambiguous children.

The middle grid (Children associated with parent) provides a list of the children that are associated with the ambiguous parent in the combined data and statistics on their occurrence (percentage of sites and samples where child occurs) and the percentage of the total of the children's

abundances (or densities) contributed by each child. The Options menu (fig. 2) on the Data Preparation module window (fig. 39) controls the order in which children are listed in this grid. Children can be listed alphabetically or in order of occurrence based on sites. Listing children by occurrence (most common to least common) is the default setting and the setting that is most useful for selecting representative taxa (that is, taxa that have the highest probability of occurring at a site). The statistics presented here allow the user to decide on an action to take based on the abundance and occurrence of the children and parent in the dataset. For example, the user can select which children to divide the parent's abundance among by clicking on the appropriate row of the middle grid. The first time the user clicks on a row, an "X" appears in the Select column and the child is transferred to the bottom grid (Replace parent with the following children and abundances). Clicking on a row in the middle grid that has already been selected will cancel the selection. This removes the "X" from the Select column and removes the child from the bottom grid. The Children associated with parent grid can only display three children at a time. The presence of a scroll bar on the right-hand side of this grid (fig. 46) indicates that there are children that are not currently visible in this window. Click on the scroll bar to view and select other children.

The example in figure 46 shows that the user has elected to replace the ambiguous parent with the three children (*Ferrissia* sp., Pulmonata, and *Physella* sp.) visible in the bottom grid (Replace parent with the following children and abundances). Both occurrence (percentage of sites and samples where taxon occurs) and dominance (percentage of total abundance or density contributed by the taxon) should be considered when selecting children to replace the ambiguous parent. The most conservative approach is to select the dominant child, which is the child that occurs at the most sites and represents the majority of abundance or density. Using this criterion, *Ferrissia* sp. should be substituted for Gastropoda because it has the highest occurrence (19.6 percent of sites, 14.1 percent of samples) and dominance (88.7 percent of total density). If a single dominant taxon does not exist, then the analyst can select children that collectively occur at a large number of sites and represent a substantial portion of abundance and density. Taxa that occur at only a small number of sites or samples or constitute only a small proportion of abundance (for example, Pulmonata and *Physella* sp.) probably should not be included unless there is an ecological reason to

> TIP: The "Children associated with parent" grid only displays information for three children at a time. The presence of a scroll bar on the right-hand side of this window indicates that additional children can be viewed by clicking on the scroll bar.

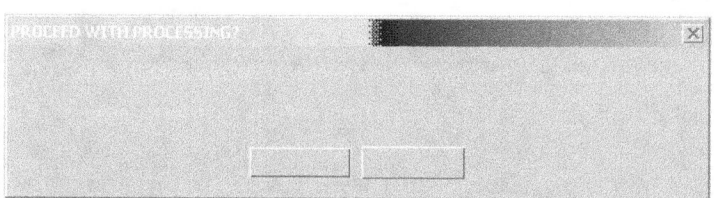

**Figure 45.** The IDAS program informs the user of the number of ambiguous taxa that need to be resolved through user intervention. The program can be exited at this time if the user decides not to manually resolve the ambiguous taxa.

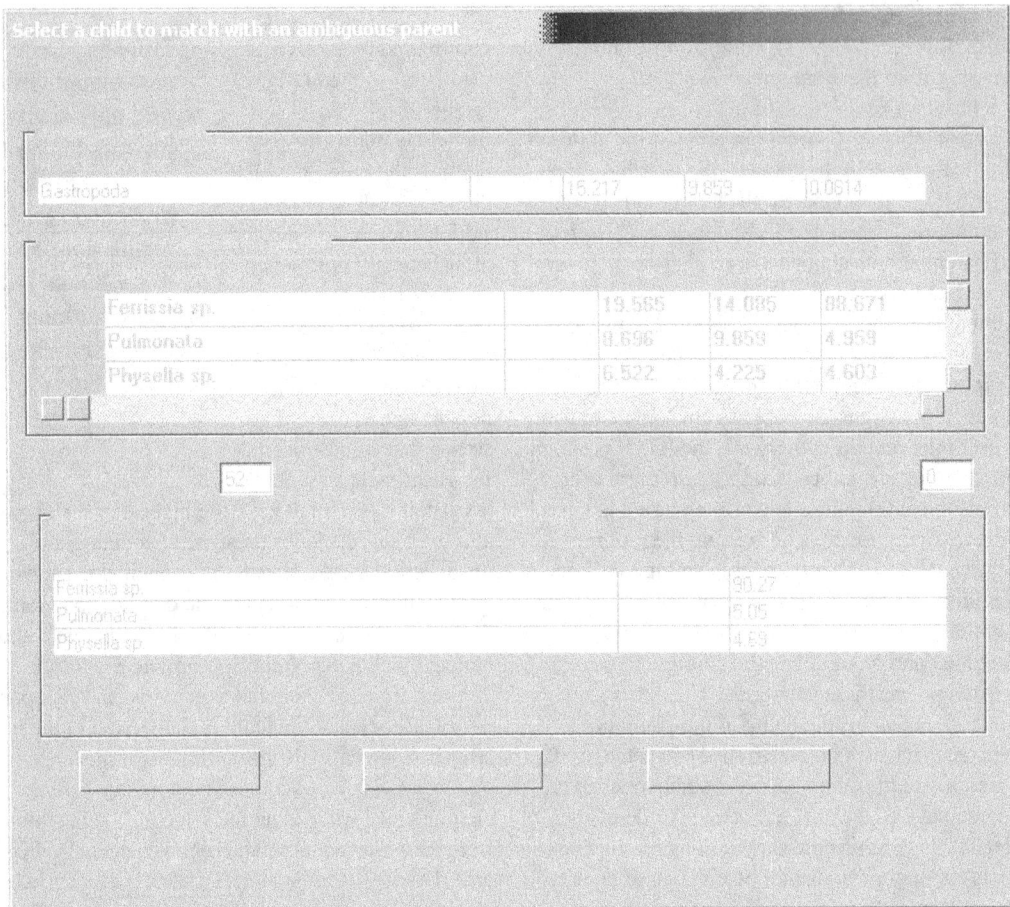

**Figure 46.**    The IDAS program allows the user to select children to match with ambiguous parents when it encounters a sample that contains an ambiguous parent but no associated children. Such situations will only be encountered when using processing methods RPKC-C and DPAC-C.

include them in the analysis. Pulmonata and *Physella* sp. were selected in figure 46 only to show the relation between the **Select** check box and the contents of the **Replace parent** grid. When making these decisions, the user is reminded that missing data are being replaced with the user's "best guess" as to what the ambiguous parent most likely would be given the distribution of ambiguous children among samples and sites. Other information, such as historical data on distributions, should be used to select the appropriate children to pair with ambiguous parents.

The IDAS program automatically calculates the percentage of the parent's abundance that will be transferred to each child as the children are transferred from the middle grid to the bottom grid. These percentages are based on the relative contribution that each of the selected children makes to total abundance or density (percentage of abundance or density in the middle grid) in the dataset formed by combining all samples together (for example, *Ferrissia* sp.—(88.671/(88.671 + 4.959 + 4.603))*100 = 90.27 percent). The user can override these values by clicking on the appropriate child in the bottom grid. A message box (fig. 47) will appear displaying the name of the taxon, current percentage (%) abundance or density

value (**Original value**) and a text box (**Revised value**) in which to enter the user-provided value. The IDAS program will check to make sure that the values of the children in the bottom grid add up to 100 before allowing the user to move on to the next ambiguous parent. If the percentages do not add up to 100, then the user will be prompted to enter new values.

Clicking on the **Accept selection(s)** button (fig. 46) will cause the IDAS program to divide the parent's abundance among the selected children in accordance with the percentages listed in the bottom grid. Selecting children in this

**Figure 47.**    This message box can be used to manually change the percentage of parent abundance that is assigned to a child.

fashion adds new taxa to each sample in which an ambiguous parent occurs without any children. The IDAS program also supports two more conservative approaches to dealing with this problem. The user can elect to delete (Delete parent button) or retain (Retain parent button) the ambiguous parent in cases where there are no associated children. It is important to remember that these three options make modifications to the data ONLY in a sample in which a taxon has been identified as an ambiguous parent but no children occur in the sample. In all other cases, the taxonomic ambiguities are handled in the same way as they would be in the case of resolving ambiguities for individual samples.

The effects of distributing, deleting, or retaining the abundances of ambiguous parents in cases where the parent is present in a sample but not the associated children are compared in table 50 for processing method RPKC-C (delete ambiguous parents, retain children) as applied to the data in table 35. If the user chooses to select children among which to distribute the abundances of ambiguous parents, then the abundance of *Baetis* sp. L (sample 7650) and *Acentrella* sp. L (sample 7973) will be distributed among the four species of *Baetis* (*B. flavistriga* L, *B. intercalaris* L, *B. pluto* L, and *B. tricaudatus* L) and two species of *Acentrella* (*A. parvula* L and *A. turbida* L) selected by the user (table 50A). This distribution will be in accordance with the abundance of each taxon in the combined dataset. For example, the proportion of *Baetis* sp. L abundance (35) distributed to *Baetis flavistriga* L in sample 7650 is the abundance of *Baetis flavistriga* L (60) in the combined dataset divided by the sum of the abundances of the selected children (60/[60+99+58+13]=0.26) in the combined dataset (table 35). The abundance of *Baetis flavistriga* L becomes the abundance of *Baetis* sp. L (35) multiplied by this proportion (0.2609) or 9.13. Similarly, the abundance of *Acentrella* sp. L in sample 7973 (30) is divided between *Acentrella parvula* L (30*(71/[71+12] )= 25.66) and *Acentrella turbida* L. (30*(12/[71+12])=4.34).

Ambiguous parents that have children in the dataset (Ephemeroptera L [100], *Plauditus* sp. L [5] in sample 7650; Baetidae L [8], *Baetis* sp. L [100], Zygoptera L [100] in sample 7896; Baetidae L [4], *Acentrella* sp. L [10], *Pseudocloeon* sp. L [1] in sample 7729; and *Baetis* sp. L [44] in sample 7973) are deleted in method RPKC-C, and their abundances are removed from the samples (110 in sample 7650, 208 in sample 7896, 15 in sample 7729, and 44 in sample 7973). Ephemeroptera A and *Diplectrona* sp. P are not identified as ambiguous taxa (table 35), because the lifestage information associated with these taxa makes them unique. If lifestage information had not been retained, then Ephemeroptera A and Hydropsychidae L would have been dropped from the dataset.

If the user chooses to delete ambiguous parents in samples where no associated children are present (table 50B), then *Baetis* sp. L in sample 7650 and *Acentrella* sp. L in sample 7973 would be dropped. It is important for the analyst to note that the IDAS program will automatically substitute children for parents in cases where the parent can only be associated with a single child (for example, *Pseudocloeon* sp. L and Zygoptera L in sample 7650; *Plauditus* sp. L in sample 7896). The RPKC-C method also will delete ambiguous parents that have children in the samples (Ephemeroptera L, Baetidae L, *Plauditus* sp. L, Zygoptera L, *Acentrella* sp. L, and *Pseudocloeon* sp. L). The difference between the Delete option and the Distribute option is the loss of all *Baetis* mayflies in sample 7650 and *Acentrella* mayflies in sample 7973, which has a substantial effect on taxa richness and total abundance.

If the user chooses to retain ambiguous parents in samples where no children are present (table 50C) by using method RPKC-C, then *Acentrella* sp. L is retained in sample 7973 and *Baetis* sp. L in sample 7650. Ambiguous parents that have children in the samples are deleted. The result is an increase in taxa richness in samples 7650 and 7973 and total abundances that are similar to those obtained when the abundances are distributed among the children. The important thing to remember is that using this option to retain ambiguous parents will result in datasets that still contain ambiguous taxa.

The effects of distributing, deleting, or retaining ambiguous parents in samples without associated children are similar for processing method DPAC-C (table 51) to what was observed for processing method RPKC-C (table 50). The same taxa are deleted or retained; however, method DPAC-C distributes the abundances of ambiguous parents among the associated children rather than discarding these abundances as is done in method RPKC-C. As with method RPKC-C, the IDAS program automatically substitutes the child for an ambiguous parent in cases where there is only one child associated with the parent (for example, Zygoptera L and *Argia* sp. L; *Pseudocloeon* sp. L and *Pseudocloeon propinquum* L; *Plauditus* sp. L and *Plauditus cestus* L).

The "distribute" option in method DPAC-C distributes the parent's abundance among the children based on the relative abundance of the children in the sample or the relative abundances of the children selected by the user in those cases where no children are present in the sample (table 51B). As with method RPKC-C, the relative abundances are based on the abundances of the children in the combined samples (table 35). The user can override the values ("% abundance" column in the Replace parent with the following children and abundances grid, fig. 46) assigned to the children by clicking on each of the selected children and assigning new values (fig. 47). Abundances are distributed iteratively starting at the genus level and progressing up to phylum.

If the user elects to delete ambiguous parents in cases where there are no associated children in the sample, the IDAS program will delete the parent and its abundance before distributing the abundance of any remaining ambiguous parents among the associated children (table 51B). Similarly, if the user chooses to retain ambiguous parents in samples where there are no associated children, the IDAS program will distribute the abundance of ambiguous parents while taking the retained parents' abundances into consideration. This will result in a processed dataset that contains taxonomic ambiguities (table 51C).

**Table 50.**    Effects of distributing, deleting, or retaining parents when processing ambiguous taxa by using method RPKC-C.

[In situations where ambiguous parents exist in a sample but no ambiguous children exist, the user can (A) distribute abundance of ambiguous parents among children, (B) delete ambiguous parents, or (C) retain ambiguous parents. These methods are illustrated by applying them to the data presented in table 35]

(A) Distribute ambiguous parents abundance among children. *Acentrella* sp. L was distributed between *A. turbida* (14.5 %) and *A. parvula* (85.5 %). Baetidae was distributed among *B. flavistriga* L (26.09 %), *B. intercalaris* L (43.09 %), *B. pluto* L (25.22 %), and *B. tricaudatus* L (5.65 %). Abundances of *Plauditus* sp. L, *Pseudocloeon* sp. L, and Zygoptera L are automatically assigned to *Plauditus cestus* L, *Pseudocloeon propinquum* L, and *Argia* sp. L because only one child is associated with each of these taxa.

| Taxon | Sample | | | |
|---|---|---|---|---|
| | 7650 | 7896 | 7729 | 7973 |
| Ephemeroptera A. | 2.00 | | | |
| *Acentrella parvula* L | 26.00 | 11.00 | 34.00 | 25.66 |
| *Acentrella turbida* L | | 9.00 | 3.00 | 4.34 |
| *Baetis flavistriga* L | 9.13 | 6.00 | 54.00 | |
| *Baetis intercalaris* L | 15.06 | | 78.00 | 21.00 |
| *Baetis pluto* L | 8.83 | 12.00 | 23.00 | 23.00 |
| *Baetis tricaudatus* L | 1.98 | 2.00 | 11.00 | |
| *Plauditus cestus* L | 34.00 | 56.00 | | 53.00 |
| *Pseudocloeon propinquum* L | 45.00 | | 5.00 | 112.00 |
| *Argia* sp. L | 10.00 | 8.00 | | |
| Hydropsychidae L | 50.00 | | | |
| *Diplectrona* sp. A | 10.00 | | 23.00 | |
| Richness | 11 | 7 | 8 | 6 |
| Total abundance | 212.00 | 104.00 | 231.00 | 239.00 |

(B) Delete ambiguous parents when no children are present in sample. *Acentrella* sp. L and *Baetis* sp. L are deleted from samples 7650 and 7973, respectively, and their abundances are lost. Abundances of *Plauditus* sp. L, *Pseudocloeon* sp. L, and Zygoptera L are automatically assigned to *Plauditus cestus* L, *Pseudocloeon propinquum* L, and *Argia* sp. L because only one child is associated with each of these taxa.

| Taxon | Sample | | | |
|---|---|---|---|---|
| | 7650 | 7896 | 7729 | 7973 |
| Ephemeroptera A | 2.00 | | | |
| *Acentrella parvula* L | 26.00 | 11.00 | 34.00 | |
| *Acentrella turbida* L | | 9.00 | 3.00 | |
| *Baetis flavistriga* L | | 6.00 | 54.00 | |
| *Baetis intercalaris* L | | | 78.00 | 21.00 |
| *Baetis pluto* L | | 12.00 | 23.00 | 23.00 |
| *Baetis tricaudatus* L | | 2.00 | 11.00 | |
| *Plauditus cestus* L | 34.00 | 56.00 | | 53.00 |
| *Pseudocloeon propinquum* L | 45.00 | | 5.00 | 112.00 |
| *Argia* sp. L | 10.00 | 8.00 | | |
| Hydropsychidae L | 50.00 | | | |
| *Diplectrona* sp. A | 10.00 | | 23.00 | |
| Richness | 7 | 7 | 8 | 4 |
| Total abundance | 177.00 | 104.00 | 231.00 | 209.00 |

**Table 50.**   Effects of distributing, deleting, or retaining parents when processing ambiguous taxa by using method RPKC-C—Continued

[In situations where ambiguous parents exist in a sample but no ambiguous children exist, the user can (A) distribute abundance of ambiguous parents among children, (B) delete ambiguous parents, or (C) retain ambiguous parents. These methods are illustrated by applying them to the data presented in table 35]

(C) Retain ambiguous parents when no children are present in a sample; otherwise, distribute abundances among children. *Acentrella* sp. L and *Baetis* sp. L are retained. Abundances of *Plauditus* sp. L, *Pseudocloeon* sp. L, and Zygoptera L are automatically assigned to *Plauditus cestus* L, *Pseudocloeon propinquum* L, and *Argia* sp. L because only one child is associated with each of these taxa.

| Taxon | Sample | | | |
|---|---|---|---|---|
| | 7650 | 7896 | 7729 | 7973 |
| Ephemeroptera A | 2.00 | | | |
| *Acentrella* sp. L | | | | 30.00 |
| *Acentrella parvula* L | 26.00 | 11.00 | 34.00 | |
| *Acentrella turbida* L | | 9.00 | 3.00 | |
| *Baetis* sp. L. | 35.00 | | | |
| *Baetis flavistriga* L | | 6.00 | 54.00 | |
| *Baetis intercalaris* L | | | 78.00 | 21.00 |
| *Baetis pluto* L | | 12.00 | 23.00 | 23.00 |
| *Baetis tricaudatus* | | 2.00 | 11.00 | |
| *Plauditus cestus* | 34.00 | 56.00 | | 53.00 |
| *Pseudocloeon propinquum* | 45.00 | | 5.00 | 112.00 |
| *Argia* sp. L | 10.00 | 8.00 | | |
| Hydropsychidae L | 50.00 | | | |
| *Diplectrona* sp. P | 10.00 | | | |
| Richness | 8 | 7 | 8 | 5 |
| Total abundance | 212.00 | 104.00 | 231.00 | 239.00 |

**Table 51.** Effects of distributing, deleting, or retaining parents when processing ambiguous taxa by using method DPAC-C.

[In situations where ambiguous parents exist in a sample but no ambiguous children exist, the user can (A) choose children among whom the parent's abundance will be distributed, (B) delete ambiguous parents, or (C) retain ambiguous parents. These methods are illustrated by applying them to the data presented in table 35]

(A) Distribute ambiguous parent abundance among children. *Acentrella* sp. L was distributed between *A. turbida* (14.5 %) and *A. parvula* (85.5 %). Baetidae was distributed among *B. flavistriga* L (26.09 %), *B. intercalaris* L (43.09 %), *B. pluto* L (25.22 %), and *B. tricaudatus* L (5.65 %). Abundances of *Plauditus* sp. L, *Pseudocloeon* sp. L, and Zygoptera L are automatically assigned to *Plauditus cestus* L, *Pseudocloeon propinquum* L, and *Argia* sp. L because only one child is associated with each of these taxa.

| Taxon | Sample | | | |
|---|---|---|---|---|
| | 7650 | 7896 | 7729 | 7973 |
| Ephemeroptera A. | 2.00 | | | |
| *Acentrella parvula* L | 51.67 | 11.45 | 43.98 | 25.67 |
| *Acentrella turbida* L | | 9.37 | 3.88 | 4.34 |
| *Baetis flavistriga* L | 15.22 | 37.47 | 54.99 | |
| *Baetis intercalaris* L | 25.11 | | 79.42 | 42.00 |
| *Baetis pluto* L | 14.71 | 74.94 | 23.42 | 46.00 |
| *Baetis tricaudatus* L | 3.30 | 12.49 | 11.20 | |
| *Plauditus cestus* L | 65.00 | 58.29 | | 53.00 |
| *Pseudocloeon propinquum* L | 75.00 | | 6.11 | 112.00 |
| *Argia* sp. L | 10.00 | 108.00 | | |
| Hydropsychidae L | 50.00 | | | |
| *Diplectrona* sp. A | 10.00 | | 23.00 | |
| Richness | 11 | 7 | 8 | 6 |
| Total abundance | 322.00 | 312.00 | 246.00 | 283.00 |

(B) Delete parents when no children are present in sample. *Acentrella* sp. L and *Baetis* sp. L are deleted from samples 7650 and 7973, respectively, and their abundances are lost. Abundances of *Plauditus* sp. L, *Pseudocloeon* sp. L, and Zygoptera L are automatically assigned to *Plauditus cestus* L, *Pseudocloeon propinquum* L, and *Argia* sp. L because only one child is associated with each of these taxa.

| Taxon | Sample | | | |
|---|---|---|---|---|
| | 7650 | 7896 | 7729 | 7973 |
| Ephemeroptera A | 2.00 | | | |
| *Acentrella parvula* L | 57.96 | 11.45 | 43.98 | |
| *Acentrella turbida* L | | 9.37 | 3.88 | |
| *Baetis flavistriga* L | | 37.47 | 54.99 | |
| *Baetis intercalaris* L | | | 79.42 | 42.00 |
| *Baetis pluto* L | | 74.94 | 23.42 | 46.00 |
| *Baetis tricaudatus* L | | 12.49 | 11.20 | |
| *Plauditus cestus* L | 72.91 | 58.29 | | 53.00 |
| *Pseudocloeon propinquum* L | 84.13 | | 6.11 | 112.00 |
| *Argia* sp. L | 10.00 | 108.00 | | |
| Hydropsychidae L | 50.00 | | | |
| *Diplectrona* sp. A | 10.00 | | 23.00 | |
| Richness | 7 | 7 | 8 | 4 |
| Total abundance | 287.00 | 312.00 | 246.00 | 253.00 |

**Table 51.** Effects of distributing, deleting, or retaining parents when processing ambiguous taxa by using method DPAC-C—Continued

(C) Retain parents when no children are present in a sample, otherwise distribute abundances among children. *Acentrella* sp. L and *Baetis* sp. L are retained. Abundances of *Plauditus* sp. L, *Pseudocloeon* sp. L, and Zygoptera L are automatically assigned to *Plauditus cestus* L, *Pseudocloeon propinquum* L, and *Argia* sp. L because only one child is associated with each of these taxa.

| Taxon | Sample | | | |
|---|---|---|---|---|
| | 7650 | 7896 | 7729 | 7973 |
| Ephemeroptera A | 2.00 | | | |
| *Acentrella* sp. L | | | | 30.00 |
| *Acentrella parvula* L | 51.67 | 11.45 | 43.98 | |
| *Acentrella turbida* L | | 9.37 | 3.88 | |
| *Baetis* sp. L | 58.33 | | | |
| *Baetis flavistriga* L | | 37.47 | 54.99 | |
| *Baetis intercalaris* L | | | 79.42 | 42.00 |
| *Baetis pluto* L | | 74.94 | 23.42 | 46.00 |
| *Baetis tricaudatus* | | 12.49 | 11.20 | |
| *Plauditus cestus* | 65.00 | 58.29 | | 53.00 |
| *Pseudocloeon propinquum* | 75.00 | | 6.11 | 112.00 |
| *Argia* sp. L | 10.00 | 108.00 | | |
| Hydropsychidae L | 50.00 | | | |
| *Diplectrona* sp. P | 10.00 | | 23.00 | |
| Richness | 8 | 7 | 8 | 5 |
| Total abundance | 322.00 | 312.00 | 246.00 | 283.00 |

The examples presented in tables 50 and 51 did not mix the Distribute, Delete, or Retain options in order to simplify these examples. However, the IDAS program allows the analyst to mix these options when dealing with ambiguous parents. These actions will be applied only to the ambiguous parents that are present in a sample without associated children. The combination of these options provides the analyst with a diverse and powerful set of tools for resolving taxonomic ambiguities.

## Running the Data Preparation Module

Once the user has made selections on the Data Preparation module's main form (fig. 39), the choices can be executed by clicking on Run on the menu line at the top of the form. This action calls up the Processing status window (fig. 48), which informs the user of actions the module is currently conducting (highlighted lines), actions that have been completed (overstrike lines), and actions that have not been selected (dimmed lines). At the bottom on the Processing status window is a list box that displays the samples that have been processed (SUID, STAID, and SMCOD) and a text box that shows the percentage of samples that have been processed.

Sometimes the time interval (±1–7 days) associated with the formation of a QUAL sample will encompass multiple RTH or DTH samples. If this happens, a pop-up window will allow the user to select the appropriate RTH and (or) SMCODs to associate with the QMH SMCOD (fig. 49). The Hide and Reset menu items at the top of the Processing status window allow the user to hide the window or to reset the contents of the window. If the Processing status window is hidden, it can be reactivated by clicking on the Status menu item in the main window of the Data Preparation module (fig. 39).

When processing has been completed, the Processing status window is displayed over the Data Preparation window (fig. 39). Clicking on the Hide menu item of the Processing status window returns the user to the Data Preparation window with all the processing options preserved. A new set of sample-processing options can be applied to the open data file simply by selecting a new set of options. These options are executed, and a new "processed" data file is created by clicking on Run in the menu line. In this fashion, the user can process quantitative samples and then qualitative samples or investigate different methods of resolving taxonomic ambiguities, deleting rare taxa, or the effects of using different taxonomic levels without having to re-open the data file for each analysis.

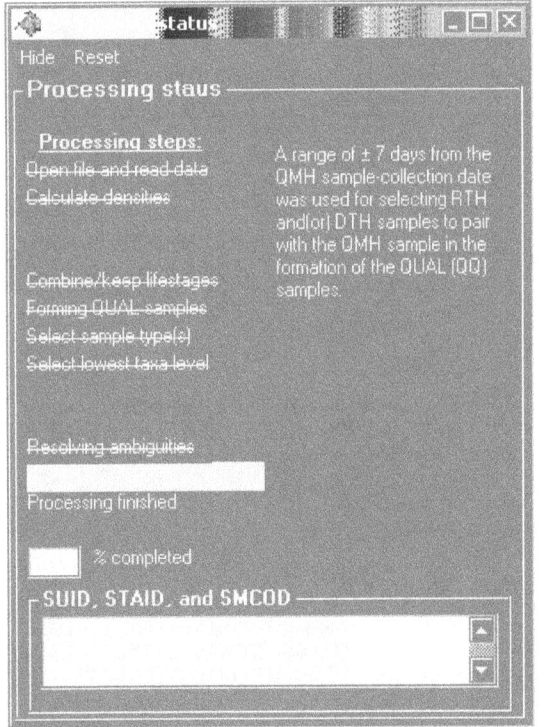

**Figure 48.** The Processing status window from the Data Preparation module.

[The highlighted line indicates that the IDAS program is currently copying results to the output files. Overstrike lines are actions that have been completed. The dimmed lines are processing options that have not been selected]

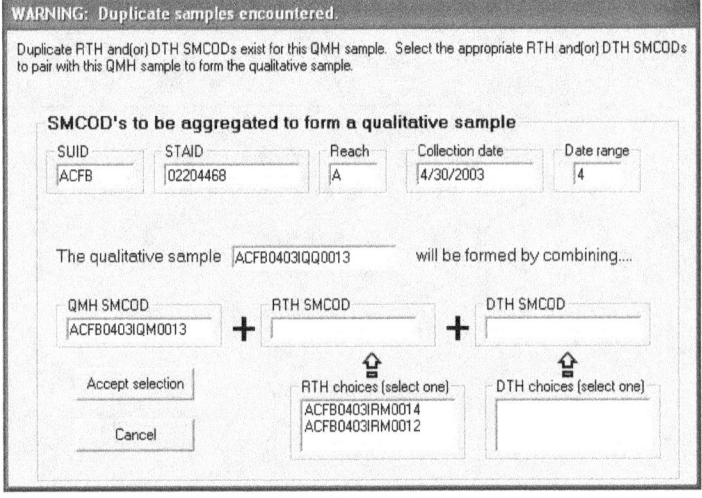

**Figure 49.** A pop-up window warns the user that multiple RTH and (or) DTH samples are associated with the QMH sample. The user must select a single RTH and DTH sample to combine with the QMH sample to form the QUAL sample.

TIP: Click on "Hide" to return to the "Data Preparation" window. You can then select new processing options to apply to the open dataset.

## Output From the Data Preparation Module

The IDAS program prompts the user to supply a name for the processed data that are stored by the Data Preparation module as a new data table or spreadsheet (fig. 14) in the Access database or Excel workbook that provided the abundance data. The name must be 15 characters or less to accommodate the suffixes (up to 12 characters in length) that IDAS uses to identify output files while still meeting the 31-character limit for spreadsheet names in Excel. The default name that the IDAS program provides for the processed data is "NoAmbig." The format of the processed data (table 3) is slightly different from that of the original data (table 1) and is arranged in order by SUID, STAID, Reach, CollectionDate, SampleID, SortCode, and Lifestage.

In addition to the processed data, this module will, depending on the processing options selected by the user, produce another two or three tables or spreadsheets that document the data-processing steps selected by the user. These tables are stored in the database or workbook that provided the abundance data. The options table is stored under the user-supplied name plus the suffix "_Options" (for example, NoAmbig_Options). The options table documents the data preparation by storing information on the name of the source data file(s), the output data table or spreadsheet, and the options selected (table 52).

> **TIP: Output data tables and spreadsheets are automatically appended to the workbook or database that provided the data.**

**Table 52.** The options selected by the user in the Data Preparation module are stored in a spreadsheet or data table with the suffix "_Options."

[The options file makes IDAS self-documenting, which facilitates the documentation and archiving of data-analysis procedures]

| Options | Selected |
| --- | --- |
| Data Prep module run: | 7/15/2009 at 11:17:40 AM |
| IDAS version: | v. 5.0.0 |
| Source file for abundance data | D:\IDAS_20090715_1104_tcuffney.mdb |
| Source table/spreadsheet for abundance data: | Invert_Results_Comb |
| Calculate densities | Yes |
| Source file for area sampled | D:\IDAS_20090715_1104_tcuffney.mdb |
| Source table/spreadsheet for area sampled | Sample_All |
| Sample type(s) selected for processing | All |
| Delete artifacts | ALL |
| Delete immatures | AMBIG |
| Delete damaged specimens | ALL |
| Delete specimens with wrong gender for identification | ALL |
| Delete specimens with indeterminate identifications | AMBIG |
| Delete specimens where poor mounts interfere with IDs | AMBIG |
| Delete pupae | No |
| Delete terrestrial adults | Yes |
| Keep lifestages separate | No |
| Combine lifestages for each BU_ID | Yes |
| QUAL sample formed from | QMH+RTH+DTH |
| Range allowed in collection dates (days) to form QUAL sample | 7 |
| Lowest taxonomic level allowed | Species |
| Delete rare taxa simultaneously | Delete taxa if they are found at three sites or less ***AND*** they constitute 0.01% of sample abundance or less. |
| Before resolving ambiguities delete ambiguous parents at or above: | >= Class |
| Keep samples separate while resolving ambiguities | Yes |
| Option 1: Drop parents, keep children | No |
| Option 2: Add children to parents | No |
| Option 3: Keep children if greater than parents | No |
| Option 4: Distribute parents among children | Yes |
| Option 5: None — do not resolve ambiguities | No |

Statistics on the number of rows of data and abundance associated with each sample are calculated at three points in the data-processing stream. These statistics are stored in a table or spreadsheet named by combining the user-supplied name plus the suffix "_Stats" (for example, NoAmbig_Stats). An example of the data-preparation statistics that are output by the IDAS program is given in table 53. For each sample, the numbers of rows are given for (1) the original data (oRows); (2) after processing laboratory notes, lifestages, selecting lowest taxa levels, and deleting rare taxa (dRows); and (3) after resolving ambiguities (aRows). Total abundances at each processing stage are stored in the columns oAbund, dAbund, and aAbund (or oDensity, dDensity, and aDensity—if the "calculate density" option was chosen). These statistics give insight into how much the information content of the data changes as these samples are prepared for analysis.

If the user chooses to form QUAL samples from compilations of RTH, DTH, and QMH samples, the IDAS program will store information on which RTH and DTH samples are paired with QMH samples. It will also store a list of the SampleIDs and SMCODs that are created to identify the QUAL sample. All of this information is stored in a new spreadsheet or data table that is named by combining the user-supplied name with the suffix "_QQsmcods" (for

example, QUAL_QQsmcods; table 54). Similar information is created when COMB samples are created when processing data in WR-EMAP format. This information is stored in a new spreadsheet or data table with the suffix "_CCsmcods" (for example, NoAmbig_CCsmcods).

## Resetting or Exiting the Module

The user can reset the module by selecting the Close option from the Files menu. This will return the user to the opening window of the module and prepare the module to open a new dataset. This procedure should be followed if the user wishes to process multiple datasets through the same module. The user can exit the module by selecting Exit from the menu bar. This will close the module and return the user to the opening screen of the IDAS program (fig. 1). Alternatively, the user can click on the "x" in the upper right-hand corner of the window. To apply different processing options to the same dataset, the user does not have to close and re-open the file for processing but simply can click on the Hide menu item on the Processing status window and then select the desired processing options on the Data Preparation window and click on Run to process the data.

**Table 53.**    The IDAS program provides the user with information on the number of rows and total abundance that are present at three stages of processing in the Data Preparation module. This information is stored in a spreadsheet or table that ends with the suffix "_Stats."

[This file contains information on the number of rows (oRows) and total abundance or density (oAbund or oDensity) in the original data, the number of rows (dRows) and total abundance or density (dAbund or dDensity) after processing laboratory notes, lifestages, lowest taxonomic levels, and rare taxa, and the number of rows (aRows) and total abundance or density (aAbund or aDensity) after removing ambiguous taxa. This information can be used to assess the effects of various data-preparation methods]

| Column name | Data type | Example | Comment |
|---|---|---|---|
| SUID | Text | ALMN | Study Unit identifier |
| STAID | Text | 03015795 | Station identifier |
| Reach | Text | A | Sampling reach |
| CollectionDate | Date | 6/27/1996 | Collection date |
| SampleID | Long | 7650 | Sample identifier |
| SMCOD | Text | ALMN0696IRM0001 | Sample code |
| oRows | Long | 101 | Rows in original data |
| dRows | Long | 93 | Rows prior to resolving ambiguities |
| aRows | Long | 61 | Rows after resolving ambiguities |
| oDensity | Long | 2060 | Density in original data |
| dDensity | Long | 2060 | Density prior to resolving ambiguities |
| aDensity | Double | 1810.4 | Density after resolving ambiguities |

**Table 54.** The IDAS program documents the RTH and (or) DTH samples that are paired with QMH samples in the formation of qualitative (QUAL) samples (NoAmbig_QQsmcods).

[This information is stored in a spreadsheet or data table that ends with the suffix "_QQSMCODs." Columns "QSMCOD" and "QSampleID" contain the SMCOD and SampleID used to identify QUAL samples. These identifiers are based on the QMH sample SMCOD (QMHsmcod) and SampleID (QMHSampleID). The RTH and DTH samples that are paired with the QMH samples are identified by SMCOD (RTHsmcod and DTHsmcod) and SampleID (RTHSampleID and DTHSampleID)]

| Column name | Data type | Example | Comments |
|---|---|---|---|
| SUID | Text | ALMN | Study Unit identifier |
| STAID | Text | 03015795 | Station identifier |
| Reach | Text | A | Sampling reach |
| CollectionDate | Date | 6/28/1996 | Collection date |
| QSMCOD | Text | ALMN0696IQQ0002 | QQ sample code |
| QSampleID | Long | -7654 | QQ sample identifier |
| QMHsmcod | Text | ALMN0696IQM0002 | QMH sample code |
| QMHSampleID | Long | 7654 | QMH sample identifier |
| RTHsmcod | Text | ALMN0696IRM0001 | RTH sample code |
| RTHSampleID | Long | 7650 | RTH sample identifier |
| DTHsmcod | Text | ALMN0696IDM0001 | DTH sample code |
| DTHSampleID | Long | 7658 | DTH sample identifier |

# Calculate Community Metrics Module

The Calculate Community Metrics module calculates 187 community metrics and is started by clicking on the Calculate community metrics button on the main program window (fig. 1). This module uses the processed data format (table 3) produced by the Data Preparation module. The module will not accept data in Bio-TDB, WR-EMAP, or user-defined format because it assumes that each row in a sample corresponds to what the user has decided is a unique taxon (BU_ID or BU_ID + lifestage). This module (fig. 50) uses standard menu items for opening and closing files (Files), viewing data files (View), setting program options (Options), exiting the module (Exit), displaying information about the module (About), and for executing selected processing options (Run). The View menu also can be used to view and print a list of the

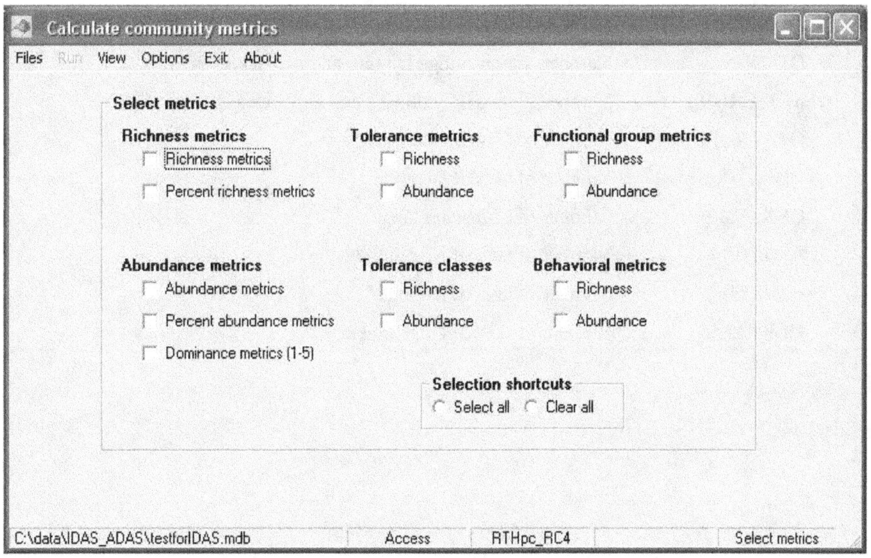

TIP: The "Calculate Community Metrics" module uses the "processed" data format produced by the "Data Preparation" module.

[The Select metrics frame appears after the user has opened an Excel spreadsheet or Access table for processing]

**Figure 50.** Main window of the Calculate Community Metrics module.

community metrics calculated by this module. The standard five-panel **status bar** located along the bottom of the module window displays (from left to right) the name of the source file, the source file type (Excel® or Access®), the name of the spreadsheet or data table that is the source of the data, the name of the spreadsheet or data table that will store the processed data, and processing status messages.

The Calculate Community Metrics module derives metrics based on taxa richness (table 55), abundance (table 56), functional groups (table 57), tolerance (table 58), dominance (tables 59 and 60), and behavioral groups (table 61). The metrics calculated by IDAS and the abbreviations used to identify them can be viewed and printed within IDAS by clicking on the View and List of metrics menus of the Calculate Community Metrics module.

**Table 55.** Richness (RTH_R_Metrics) and percentage (%) richness (RTH_pR_Metrics) metrics calculated by the IDAS program.

| Abbreviation | | Description |
|---|---|---|
| **Richness** | **% Richness** | |
| RICH | | Total richness (number of taxa) |
| EPTR | EPTRp | Number of Ephemeroptera, Plecoptera, and Trichoptera taxa |
| EPT_CHR | EPT_CHRp | Ratio of EPT to midge taxa (EPTR/CHR, EPTRp/CHRp) |
| EPEMR | EPEMRp | Number of Ephemeroptera taxa |
| PLECOR | PLECORp | Number of Plecoptera taxa |
| PTERYR | PTERYRp | Number of *Pteronarcys* taxa |
| TRICHR | TRICHRp | Number of Trichoptera taxa |
| ODONOR | ODONORp | Number of Odonata taxa |
| COLEOPR | COLEOPRp | Number of Coleoptera taxa |
| DIPR | DIPRp | Number of Diptera taxa |
| CHR | CHRp | Number of Chironomidae taxa |
| ORTHOR | ORTHORp | Number of Orthocladinae midge taxa |
| ORTHO_CHR | ORTHO_CHRp | Ratio of Orthocladinae to midge taxa (ORTHOR/CHR, ORTHORp/CHRp) |
| TANYR | TANYRp | Number of Tanytarsini midge taxa |
| TANY_CHR | TANY_CHRp | Ratio of Tanytarsini to midge taxa (TANYR/CHR, TANYRp/CHRp) |
| NCHDIPR | NCHDIPRp | Number of non-midge Diptera taxa |
| NONINSR | NONINSRp | Number of non-insect taxa |
| ODIPNIR | ODINIRp | Number of non-midge Diptera and non-insect taxa |
| MOLCRUR | MOLCRURp | Number of taxa of Mollusca and Crustacea |
| GASTROR | GASTRORp | Number of Gastropoda taxa |
| BIVALVR | BIVALVRp | Number of taxa of Bivalvia |
| CORBICR | CORBICRp | Number of *Cobricula* taxa |
| AMPHIR | AMPHIRp | Number of taxa of Amphipoda |
| ISOPODR | ISOPORp | Number of taxa of Isopoda |
| OLIGOR | OLIGORp | Number of taxa of Oligochaeta |

**Table 56.**    Abundance (RTH_A_Metrics) and percentage (%) abundance (RTH_pA_Metrics) metrics calculated by the IDAS program.

| Abbreviation | | Description |
|---|---|---|
| **Abundance** | **% Abundance** | |
| ABUND | | Total number of organisms in the sample |
| EPT | EPTp | Abundance Ephemeroptera, Plecoptera, and Trichoptera |
| EPT_CH | EPT_CHp | Ratio of EPT to midge abundance (EPT/CH, EPTp/CHp) |
| EPEM | EPEMp | Abundance of Ephemeroptera |
| PLECO | PLECOp | Abundance of Plecoptera |
| PTERY | PTERYp | Abundance of *Pteronarcys* |
| TRICH | TRICHp | Abundance of Trichoptera |
| ODONO | ODONOp | Abundance of Odonata |
| COLEOP | COLEOPp | Abundance of Coleoptera |
| DIP | DIPp | Abundance of Diptera |
| CH | CHp | Abundance of Chironomidae |
| ORTHO | ORTHOp | Abundance of Orthocladinae midges |
| ORTHO_CH | ORTHO_CH | Ratio of Orthocladinae to midge abundance (ORTHO/CH, ORTHOp/CHp) |
| TANY | TANYp | Abundance of Tanytarsini midges |
| TANY_CH | TANY_CHp | Ratio of Tanytarsini to midge abundance (TANY/CH, TANYp/CHp) |
| NCHDIP | NCHDIPp | Abundance of non-midge Diptera |
| NONINS | NONINSp | Abundance of non-insects |
| ODIPNI | ODIPNIp | Abundance of non-midge Diptera and non-insects |
| MOLCRU | MOLCRUp | Abundance of Mollusca and Crustacea |
| GASTRO | GASTROp | Abundance of Gastropoda |
| BIVALV | BIVALp | Abundance of Bivalvia |
| CORBIC | CORBICp | Abundance of *Corbicula* |
| AMPHI | AMPHIp | Abundance of Amphipoda |
| ISOPOD | ISOPp | Abundance of Isopoda |
| OLIGO | OLIGOp | Abundance composed of Oligochaeta |

**Table 57.** Functional group metrics (RTH_FG_Metrics) calculated by the IDAS program.

[Functional group metrics are calculated on the basis of richness, percentage (%) richness, abundance, and percentage (%) abundance]

| Abbreviation | | Description |
|---|---|---|
| Richness | % Richness | |
| PA_Rich | pPA_Rich | Parasites |
| PR_Rich | pPR_Rich | Predators |
| OM_Rich | pOM_Rich | Omnivores |
| GC_Rich | pGC_Rich | Collector-gatherers |
| FC_Rich | pFC_Rich | Filtering-collectors |
| SC_Rich | pSC_Rich | Scrapers |
| SH_Rich | pSH_Rich | Shredders |
| PI_Rich | pPI_Rich | Piercers |
| FG_RICH_class | | Percentage of taxa (total richness) assigned a tolerance value |

| Abbreviation | | Description |
|---|---|---|
| Abundance | % Abundance | |
| PA_Abund | pPA_Abund | Parasites |
| PR_Abund | pPR_Abund | Predators |
| OM_Abund | pOM_Abund | Omnivores |
| GC_Abund | pGC_Abund | Collector-gatherers |
| FC_Abund | pFC_Abund | Filtering-collectors |
| SC_Abund | pSC_Abund | Scrapers |
| SH_Abund | pSH_Abund | Shredders |
| PI_Abund | pPI_Abund | Piercers |
| FG_ABUND_class | | Percentage of total abundance assigned to a functional group |

**Table 58.** Tolerance metrics (RTH_TOL_Metrics) calculated by the IDAS program.

[Tolerance metrics are calculated for richness, percentage (%) richness, abundance, and percentage (%) abundance. The criteria that define tolerance classes (intolerant, moderately tolerant, and tolerant) are set by the user in the Calculate Community Metrics module]

| Abbreviation | | Description |
|---|---|---|
| **Richness** | **% Richness** | **Description** |
| RichTOL | | Mean of taxa tolerances: $\sum_{i=1}^{N} TV_i / N$ <br> where $TV_i$ is the tolerance value of taxon "$i$" and $N$ is the number of taxa in the sample |
| Intol_rich | Intol_richp | Intolerant class: number or percentage of taxa with $TV_i \leq X$, where X is the user-specified criteria for the intolerant class |
| Modtol_rich | Modtol_richp | Moderately tolerant class: number or percentage of taxa with $X < TV_i < Y$, where X and Y are the user-specified criteria for the intolerant and tolerant classes, respectively |
| Tol_rich | Tol_richp | Tolerant class: number or percentage of taxa with $TV_i \geq Y$, where Y is the user-specified criteria for the intolerant class) |
| RICH_TOL_CLASS | | Percentage of taxa that could be assigned to a tolerance class |

| Abbreviation | | Description |
|---|---|---|
| **Abundance** | **% Abundance** | **Description** |
| AbundTOL | | Abundance-weighted mean taxa tolerances: $\sum_{i=1}^{N} TV_i A_i / \sum_{i=1}^{N} A_i$ <br> where $TV_i$ is the tolerance value of taxon "$i$," $A_i$ is the abundance of taxon "$i$," and $N$ is the number of taxa in the sample (Hilsenhoff biotic index) |
| Intol_abund | Intol_abundp | Intolerant class: abundance or percentage abundance of taxa with $TV_i \leq X$, where X is the user-specified criteria for the intolerant class) |
| Modtol_abund | Modtol_abundp | Moderately tolerant class: abundance or percentage abundance of taxa with $X < TV_i < Y$, where X and Y are the user-specified criteria for the intolerant and tolerant classes, respectively |
| Tol_abund | Tol_abundp | Tolerant class: abundance or percentage abundance of taxa with $TV_i \geq Y$, where Y is the user-specified criteria for the intolerant class) |
| ABUND_TOL_CLASS | | Percentage of abundance that could be assigned to a tolerance class |

**Table 59.**    Dominance metrics (RTH_DOM_Metrics) calculated by the IDAS program.

[Percentage (%) abundance (DOM1-DOM5) and number of taxa (DOM1R-DOM5R) in the five dominance classes. The number of taxa in the class (1-5) should correspond to the dominance class unless there were less than five taxa in the sample]

| Abbreviation | | Description |
|---|---|---|
| % Abundance | Richness | |
| DOM1 | DOM1R | Abundance (%) and taxa of most abundant taxon |
| DOM2 | DOM2R | Abundance (%) and taxa in top 2 most abundant taxon |
| DOM3 | DOM3R | Abundance (%) and taxa in top 3 most abundant taxon |
| DOM4 | DOM4R | Abundance (%) and taxa in top 4 most abundant taxon |
| DOM5 | DOM5R | Abundance (%) and taxa in top 5 most abundant taxon |

**Table 60.**    Dominance metrics calculated by the IDAS program include a list of the five most abundant taxa and their percentage (%) abundance (RTH_DOM_Taxa).

| Abbreviation | | Description |
|---|---|---|
| Name | % Abundance | |
| TAXA1 | DOM1 | Name and abundance (%) of most abundant taxon |
| TAXA2 | DOM2 | Name and abundance (%) of second most abundant taxon |
| TAXA3 | DOM3 | Name and abundance (%) of third most abundant taxon |
| TAXA4 | DOM4 | Name and abundance (%) of fourth most abundant taxon |
| TAXA5 | DOM5 | Name and abundance (%) of fifth most abundant taxon |

**Table 61.**    Behavioral metrics (RTH_BEHAV_Metrics) calculated by the IDAS program.

[Metrics are calculated for richness, percentage (%) richness, abundance, and percentage (%) abundance]

| Abbreviation | | Description |
|---|---|---|
| Richness | % Richness | |
| cn_rich | cn_richp | Clinger taxa |
| cb_rich | cb_richp | Climber taxa |
| sp_rich | sp_richp | Sprawler taxa |
| bu_rich | bu_richp | Burrower taxa |
| sw_rich | sw_richp | Swimmer taxa |
| dv_rich | dv_richp | Diver taxa |
| sk_rich | sk_richp | Skater taxa |
| BEHAV_RICH_CLASS | | Percentage of total taxa richness that was assigned a behavioral trait |

| Abbreviation | | Description |
|---|---|---|
| Abundance | % Abundance | |
| cn_abund | cn_abundp | Clinger taxa |
| cb_abund | cb_abundp | Climber taxa |
| sp_abund | sp_abundp | Sprawler taxa |
| bu_abund | bu_abundp | Burrower taxa |
| sw_abund | sw_abundp | Swimmer taxa |
| dv_abund | dv_abundp | Diver taxa |
| sk_abund | sk_abundp | Skater taxa |
| BEHAV_ABUND_CLASS | | Percentage of total abundance that was assigned a behavioral trait |

## Processing Options

Once the user successfully opens a data file (fig. 5) and data table or spreadsheet (fig. 6) to process (Files/Open menus), the **Select metrics** frame appears on the main window of the Calculate Community Metrics module (fig. 50). This frame allows the user to select the richness metrics, abundance metrics, tolerance metrics and classes, functional group metrics, and behavioral metrics (tables 55 to 60) that will be calculated. The IDAS program automatically checks to see if the input data contain quantitative samples (RTH and (or) DTH samples; SHS, TRS, or RWS for WR-EMAP samples); if none are detected, the metrics requiring quantitative data are disabled (dimmed). If the dataset contains both quantitative (RTH, DTH) and qualitative (QUAL, QMH) samples, the user can select the quantitative metrics even though these metrics cannot be calculated for qualitative samples. When the IDAS program encounters a situation in which it cannot calculate a metric (for example, calculating a quantitative metric for a qualitative sample) it will assign a null (Access tables) or blank (Excel spreadsheets) value to that metric. Although this module can handle mixed datasets (qualitative and quantitative), it is strongly recommended that qualitative and quantitative samples be analyzed separately.

Richness and abundance metrics are derived from aggregations of abundance or richness data by using the taxonomic information contained in the dataset (for example, EPT richness is the count of rows of data in a sample where the column order contains the values "Ephemeroptera," "Plecoptera," or "Trichoptera;" EPT abundance is the sum of abundances for these orders). Tolerance, functional feeding group, and behavioral metrics are based on data derived from Appendix B of the U.S. Environmental Protection Agency's (USEPA) Rapid Bioassessment Protocol (RBP) (Barbour and others, 1999). USEPA tolerance data for the Southeast (SE_TOL) has been updated with data from the North Carolina Department of Environment and Natural Resources (North Carolina Department of Environment and Natural Resources, 2006). The IDAS program stores this information in the spreadsheet ATTRIB of the Excel file Attributes_BEHAV_v5a.xls (table 20). These files are installed in the same directory as the IDAS program during the installation of IDAS. The national tolerance values stored in ATTRIB are the mean of the regional tolerance values. The tolerance and functional group data in the ATTRIB spreadsheet can be updated using tools in the Edit Data module. The Update the ATTRIB spreadsheet submenu (Files/Maintain attribute files/Modify an existing file/) can be used to incorporate tolerance and functional group data from other sources, such as other Federal, State, and local agencies. Once the user has selected the metrics to be calculated, the Run menu will be activated and can be used to begin the process of calculating metrics.

If the user has elected to calculate tolerance metrics, the IDAS program will display the Set tolerance class ranges window (fig. 51). This window allows the user to select the range of tolerance values that define the tolerant and intolerant

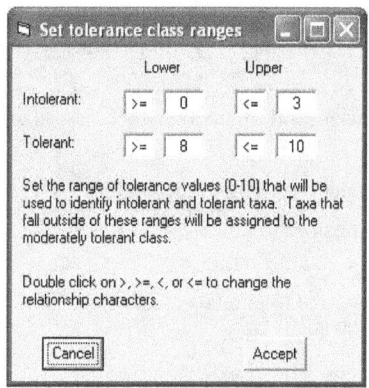

**Figure 51.** The user can select the ranges used to divide taxa into tolerant and intolerant classes.

[Taxa are assigned to the moderately tolerant class if their tolerance falls between the ranges specified for the tolerant and intolerant classes. The mathematical relations (< or <=, > or >=) can be changed by double clicking on the text box containing the symbol. Tolerance values must be between 0 and 10, and ranges may not overlap. The Options menu can be used to save the tolerance ranges as the default values for the IDAS program]

classes. Taxa with tolerance values outside of these ranges are assigned to the moderately tolerant class. The tolerance ranges can be set by entering a numeric value between 0 and 10 in the text boxes that define the tolerance value ranges. The mathematical symbols used to define the tolerance ranges can be changed by double clicking on the text box that contains the symbol to alternate between > and ≥ or < and ≤. Once the user has entered the appropriate tolerance values and made adjustments to the mathematical symbols, the Accept button is used to calculate the tolerance classes using these values. The Cancel button is used to stop processing data and reset the Calculate community metrics module. Tolerance ranges are populated with the values stored in the file IDAS_Defaults.txt when the Calculate community metrics module is started. The Options menu can be used to set the tolerance values entered into the Set tolerance class ranges window as the default values stored in IDAS_Defaults.txt (fig. 2).

Once the Accept button has been selected, the Program checks the validity of the tolerance ranges that have been entered by the user. The following situations (1–4) will generate an error message indicting that an entry in the Set tolerance class ranges window must be corrected before IDAS can continue to process data:

1. a non-numeric value is entered for the lower or upper limits

2. a value less than zero or greater than 10 is entered

3. the lower limit of a range is greater than the upper limit

4. the ranges for tolerant and intolerant classes overlap

The following situations (5–9) will generate warning messages:

5. the lower limit of the intolerant class is greater than zero

6. the upper limit of the tolerant class is less than 10

7. the range of tolerance values assigned to the moderately tolerant class is less then or equal to three

8. the range of tolerance values assigned to the intolerant class is less than or equal to two

9. the range of tolerance values assigned to the tolerant class is greater than or equal to two

The warning messages (5–9) require that the user confirm that the range settings are correct or the program will assume that the values are erroneous and return to the Set tolerance class ranges window.

If the user has elected to calculate tolerance, functional group, or behavioral metrics, the program will ask the user to select an attribute file to supply these traits. The attribute file that supplies the trait information must adhere to the format specified in table 20. This is most easily accomplished by using the Maintain attributes file tools in the Edit Data module to modify or create new files from the Attribute_BEHAV_v5.xls file provided with IDAS. Once the attribute file has been selected, the program will check to make sure that the file contains the necessary spreadsheets (EQTX, ATTRIB, HIER, VERSION, NOTES) and columns of data. The program also checks to see if the taxa (BU_ID) in the data are listed in the BU_ID column of the EQTX spreadsheet. If some taxa are missing from the EQTX spreadsheet, the Attribute file may require maintenance window will open and display a list of the taxa that could not be matched (fig. 52). The Print and Save menus can be used to send the list of taxa to a printer or save the list to a text file. The user may continue without information on the taxa listed (Continue button) or exit the program (Exit button).

The appearance of the Attribute file may require maintenance window is a good indication that the attribute file has not been optimized for the data that are being processed. Optimization of the attribute file involves updating the HIER and EQTX spreadsheets to include all the taxa in the dataset and to match tolerance values in the ATTRIB spreadsheet to taxa in the dataset for a specific region of the country. The attribute file that is provided with IDAS has not been optimized and should not be used without optimizing it for the data that are being analyzed. The attribute file can be updated manually using the information provided in the Attribute file may require maintenance window, but this is not recommended. Instead, it is highly recommended that the

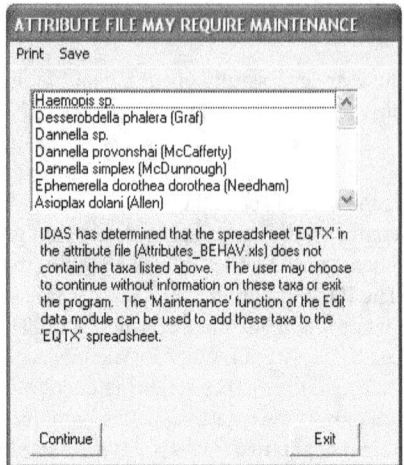

**Figure 52.** The IDAS program warns the user if it cannot match taxa with tolerance or functional group information and displays a list of names that cannot be matched. This list can be printed or saved to a text file and used to update the attribute file.

Maintain attributes file functions provided in the Edit Data module be used to update the attribute file.

The tolerance metrics are calculated as the average tolerance value of all taxa in a sample (RichTOL) or by weighting the tolerance values by the abundance of the organism in the sample (AbundTOL).

$$(1) \quad \mathrm{RichTOL} = \sum_{i=1}^{N} TV_i / N$$

$$(2) \quad \mathrm{AbundTOL} = \sum_{i=1}^{N} TV_i A_i / \sum_{i=1}^{N} A_i \quad \text{(Hilsenhoff biotic index)}$$

where

$TV_i$  is the tolerance value for taxon "$i$"
$A_i$  is the abundance of taxon "$i$" in the sample
$N$  is the number of taxa in the sample

Tolerance class metrics are calculated on the basis of how much of the taxa richness or abundance fall into the tolerance ranges specified in the Set tolerance class ranges window (fig. 51). These metrics are based on the totals (richness and abundance) for the classified taxa. That is, the percentages always add up to 100 even if the percentage of taxa richness

(RICH_TOL_class) and abundance (ABUND_TOL_class) that could be assigned to a tolerance class is less than 100.

Tolerance class range settings can be saved as the IDAS default values by using the Options menu. Clicking on this menu will call up the Set program options window (fig. 2). Click on the Save new values button in the Tolerance class range category, and the current settings will be saved to the default settings file (IDAS_Defaults.txt) once the Apply button is clicked. These settings will be automatically loaded the next time IDAS is started.

The taxa names used in the ATTRIB spreadsheet do not include authority names. Consequently, many of the BU_ID names in the Bio-TDB data file will not match entries in the ATTRIB spreadsheet of the attributes file. In addition, a taxon that lacks an attribute at the level at which it was identified (*Baetis bicaudatus*) may have an attribute at a higher taxonomic level (*Baetis* sp.). For these reasons, the IDAS program provides a taxonomic equivalency table (spreadsheet EQTX in Attributes_BEHAV_v5a.xls) to relate BU_IDs to the names listed in the ATTRIB spreadsheet (table 20). The BU_ID column in the EQTX spreadsheet contains the taxon name as it appears in the Bio-TDB data and includes the authority name. The NAME column contains the BU_ID without the authority name. The EQTXTOL and EQTXFG columns contain the equivalent taxa names for tolerance and functional groups, respectively. Equivalent taxa names are used to maximize the correspondence between the BU_IDs and the tolerance and functional feeding group data in the ATTRIB spreadsheet. The equivalent taxa are derived by looking for a direct match between the NAME field of the EQTX spreadsheet and the NAME field of the ATTRIB spreadsheet. If a direct match cannot be found, then a match is sought by setting the NAME to the next level in the taxonomic hierarchy based on the contents of the HIER spreadsheet. This process is repeated until a match is found or the level of phylum is reached. This approach maximizes the number of matches between the taxa in the data and the ATTRIB taxa lists, but it assumes that the tolerances and functional feeding groups reported at higher taxonomic levels are applicable to the aggregation of data from lower levels. The Edit Data module provides comprehensive tools (Maintain attribute files) for populating and maintaining the EQTX spreadsheet (see Attributes_BEHAV_v5a_SE_TOL.xls in Appendix II).

Once the attribute file has been opened, the Options for Tolerance and Functional Group Metrics window will be displayed (fig. 53). This window allows the user to select the method by which taxa will be matched to attributes and the regional or national tolerance values that will be used to supplement the regional tolerance values for which the EQTXTOL names have been optimized. The region that was used to optimize EQTXTOL is indicated by the regional name that is dimmed (Southeast in fig. 53). The Quit button is used to quit processing data and reset the Calculate Community

Metrics module. The Select button is used to calculate the metrics using the options selected by the user.

A taxon can be matched to tolerance (Match Tolerances using) and functional group (Match Functional Groups using) attributes in the ATTRIB spreadsheet using one of two methods (fig. 53): (1) Taxa name without authority or (2) Equivalent taxa name. The first method obtains the Name field from the EQTX spreadsheet by matching the BU_ID in the data to BU_ID field in the EQTX spreadsheet. The Name field in the EQTX spreadsheet is then matched to the Name field in the ATTRIB spreadsheet to obtain the attribute value for the taxon. The second method also matches the BU_ID in the data to the BU_ID field in the EQTX spreadsheet, but it obtains the tolerance and functional attributes by matching the names in the EQTXTOL and EQTXFG columns, respectively, to the Name field in the ATTRIB spreadsheet. The Equivalent taxa name method is the default in IDAS.

The Supplement SE_TOL with information from frame is used to select a regional or national source for tolerance values that will supplement the tolerance values for the region that was used to optimize the EQTX spreadsheet (fig. 53). For example, if the Mid-Atlantic Coastal Streams Workgroup was selected as the supplemental region, then MATL_TOL values would be used for taxa that did not have a SE_TOL tolerance value. Selecting a supplemental source for tolerance data increases the number of taxa that can be incorporated into the tolerance metrics while concentrating on the regional tolerance values. If the user wants to restrict the tolerance values to those of the region for which the EQTX spreadsheet was optimized, then the None option should be selected. This option is the default for the IDAS program.

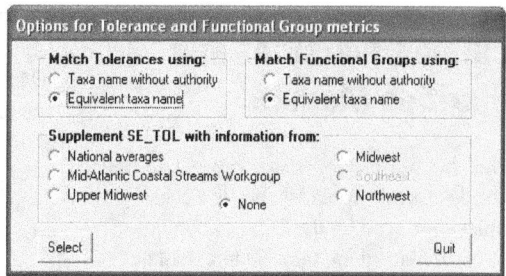

**Figure 53.** The IDAS program allows the user to determine how to match Bio-TDB taxa names to tolerance and functional group attributes. This screen also allows the user to supplement regional tolerance values (dimmed entry, Southeast in this example) with data from another region (for example, Midwest).

## Output From the Calculate Community Metrics Module

The Calculate Community Metrics module can generate up to nine new spreadsheets or tables (table 62) that are stored in the source Excel workbook or Access database. Spreadsheets or tables are named by appending suffixes to the name of the spreadsheet or data table from which the input data were derived. Each output data table or spreadsheet includes SUID, STAID, REACH, CollectionDate, SampleID, SMCOD, and the metrics as columns. The list of metrics in each output Access data table or Excel spreadsheet is given in tables 55 to 61.

## Resetting or Exiting the Module

The user can reset the module by selecting the Close option from the Files menu. This returns the user to the opening window of the module and prepares the module to accept a new dataset. This is the procedure to follow if the user wishes to process multiple datasets without exiting the module. The user can exit the module by selecting Exit from the menu bar. This closes the module and returns the user to the opening screen of the IDAS program (fig. 1). The user also can exit the module by clicking on the 'x' in the upper right-hand corner of the module window.

## Calculate Diversities and Similarities Module

The Calculate Diversities and Similarities module uses Excel® spreadsheets or Access® data tables produced by the Data Preparation module (processed data, table 3) to calculate diversity, evenness, dominance, and similarity indices (table 63). Details on the equations and procedures used to calculate these indices are given in Appendix I.

Clicking on the Calculate Diversities and Similarities button in the main program window (fig. 1) will start this module and display the Calculate diversity and similarity indices module window (fig. 54). This module uses standard menus and forms for opening and closing files (Files), viewing data files (View), executing selected processing options (Run), exiting the module (Exit), and displaying information about the module (About). The module has a standard five-panel **status bar** that displays (from left to

**Table 62.** Excel spreadsheets or Access tables that can be produced by the Calculate Community Metrics module. Names are created by assing suffixes to the source source tabe or spreadsheet name (RTH, see Appendix II).

| Name | Description |
|---|---|
| RTH_R_Metrics | Richness metrics |
| RTH_pR_Metrics | Percent richness metrics |
| RTH_A_Metrics | Abundance metrics |
| RTH_pA_Metrics | Percent abundance metrics |
| RTH_DOM_Metrics | Dominance metrics |
| RTH_DOM_Taxa | Dominant taxa and their percent abundance |
| RTH_FG_Metrics | Functional group metrics, richness and abundance |
| RTH_TOL_Metrics | Tolerance metrics, abundance, richness, and classes |
| RTH_BEHAV_Metrics | Behavior group metrics, richness and percent richness |

> **TIP: Diversity and similarity indices should be calculated separately for qualitative and quantitative samples.**

**Table 63.** Diversity and similarity indices calculated by the IDAS program.

| Diversity and evenness indices | Data requirements | Reference |
|---|---|---|
| Margalef's diversity | Quantitative | 1 |
| Menhinick's diversity | Quantitative | 1 |
| Simpson's dominance | Quantitative | 1 |
| Simpson's diversity | Quantitative | 1 |
| Brillouin's diversity | Quantitative | 1 |
| Shannon's diversity | Quantitative | 1 |
| Simpson's evenness | Quantitative | 1 |
| Brillouin's evenness | Quantitative | 1 |
| Shannon's evenness | Quantitative | 1 |

| Similarity indices | Data requirements | Reference |
|---|---|---|
| Jaccard coefficient | Qualitative | 1 |
| Sørenson coefficient | Qualitative | 1 |
| Proportional similarity | Quantitative | 1 |
| Euclidean distance | Quantitative | 1 |
| Morisita's index | Quantitative | 1 |
| Horn's index | Quantitative | 1 |
| Bray-Curtis dissimilarity | Quantitative | 2 |
| Pinkham and Pearson's index | Quantitative | 2 |

[1] Brower, J.E., and Zar, J.H., 1984, Field and laboratory methods for general ecology (2d ed.): Dubuque, IA, Wm. C. Brown Publishers, p. 226.

[2] Washington, H.G., 1984, Diversity, biotic and similarity indices. A review with special relevance to aquatic ecosystems: Water Research v. 18, no. 6, p. 653–694.

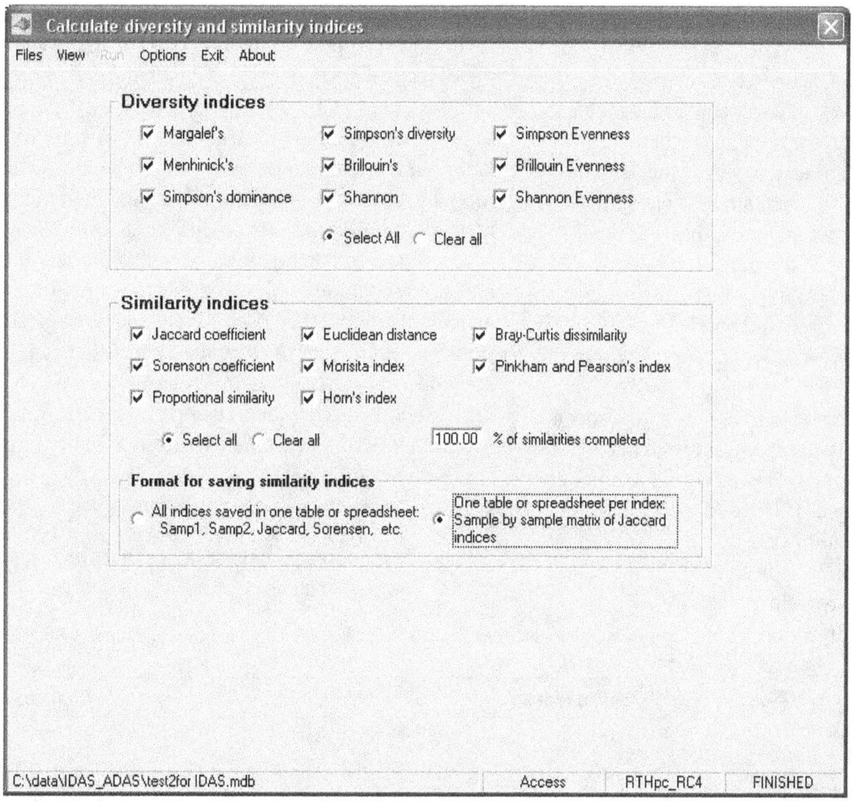

[The **Diversity indices** and **Similarity indices** frames are displayed after a data file is opened]

**Figure 54.** Main window of the Calculate Diversities and Similarities module.

right) the name of the source file, the source-file type (Excel or Access), the name of the spreadsheet or data table that is the source of the data, the name of the spreadsheet or data table that will store the processed data, and processing status information.

## Processing Options

Once a spreadsheet or data table has been successfully opened by using the Files/Open menus, the Diversity indices and Similarity indices selection frames are displayed (fig. 54). If the source dataset consists only of qualitative samples (that is, QMH or QUAL samples), the IDAS program will disable (dim) all indices that require quantitative data (that is, number of individuals in each taxon). This includes all diversity indices and all similarity indices except the Jaccard and Sørensen coefficients of similarity. If the dataset consists of mixed qualitative and quantitative samples, the IDAS program outputs blank (Excel) or null (Access) values for all quantitative indices whenever one or both of the samples being compared is qualitative. This can result in data output tables or spreadsheets that are very hard to read because of the large number of blank entries. It is recommended that diversity and similarity indices be calculated separately for qualitative and quantitative samples. The diversity and similarity indices can be selected individually by clicking on the name of the index, or all of the indices can be selected by clicking on the Select

all check box. The selections can be cleared by clicking on the name of the index or clicking on the Clear all check box.

The Format for saving similarity indices frame allows the user to select how the similarity indices will be output. All selected similarity indices can be saved in one spreadsheet (or table) or the indices can be saved in separate spreadsheets (or tables). If all indices are saved in one spreadsheet, each index becomes a column in the spreadsheet and each row corresponds to a pair of samples. If each similarity index is saved to a separate spreadsheet, each row and column corresponds to a pair of samples. The option to save similarity indices to individual spreadsheets or tables is constrained by the 255 column limit of Excel version 2003 and Access. Consequently, if the number of samples is greater than 243, a warning message is displayed and the option to save indices to separate files is disabled. Similarly, if the data originate from an Excel file with more than 362 samples, the number of similarity indices generated will exceed the approximately 65,500 row limit of Excel. When this happens, IDAS will warn the user and suggest converting the Excel file to an Access database.

> **TIP: The Calculate Diversities and Similarities module uses the "processed" data format produced by the Data Preparation module.**

## Output From the Module

Output from the Calculate Diversity and Similarity indicies module is stored in new spreadsheets or data tables within the Excel workbook or Access database that was the source for the input data. These new spreadsheets or data tables are named by appending an appropriate suffix to the name of the spreadsheet or data table that was the source of the data. For example, if the data originate from an Excel spreadsheet called "NoAmbig," then the diversity and similarity indices will be stored as the new worksheets "NoAmbig_Similarity" and "NoAmbig_Diversity" within the data file (for example, IDAS_20090715_1104_tcuffney.mdb). This module can produce from one to nine new spreadsheets within an Excel workbook or tables in an Access database depending on the options chosen by the user and the number of samples in the dataset (table 64). The IDAS program is self-documenting and will store information on the files, tables, spreadsheets, date, time, and program settings used to calculate the diversity and similarity indices in the "_Options" spreadsheet or table within the source workbook or database.

Diversity, evenness, and dominance indices are all stored in one spreadsheet or data table (table 65). In contrast, the user has the option of saving similarity indices in a single spreadsheet (or table) or to separate spreadsheets (or tables). When similarity indices are stored in a single table, each row corresponds to a couplet of samples and the columns correspond to the similarity indices selected by the user (table 66). This method of presentation is useful when the analyst is most interested in comparing differences among similarity indices rather than comparing differences in similarities among samples. Because each column corresponds to a similarity index and each row to a two-sample comparison, it is very easy to pull these data into a spreadsheet and calculate statistics that compare similarity indices (columns). However, having the rows correspond to pairs of samples makes it difficult to determine how similar samples are to one another. Saving each similarity index to a separate table produces a symmetric matrix

of similarity values in which each column compares a sample to all other samples (table 67). This format is very useful if the objective is to understand how similar samples are to one another rather than how different similarity indices compare to one another. However, if the matrix of similarities contains large numbers of blank values, which happens when analyzing mixed quantitative and qualitative data, then the symmetric matrix of similarities can be difficult to view. Therefore, it is recommended that qualitative and quantitative samples be analyzed separately unless there is a compelling reason to analyze them together. Also, Excel and Access have limitations on the number of columns of data that can be stored in a spreadsheet or data table. It is relatively easy to exceed these limits (about 256 columns) using this format in which case the Program will not allow the user to save indices to individual spreadsheets or tables.

**Table 64.** Naming conventions for spreadsheets or tables that are produced by the Calculate Diversities and Similarities module.

[This example assumes that the data originated from a spreadsheet or table named "RTH" (see Appendix II)]

| Type of index | Name | Contents |
|---|---|---|
| Diversity indices | RTH_Diversity | All diversity indices |
| Similarity indices | RTH_Similarity | All similarity indices |
| | RTH_Jaccard | Jaccard coefficient |
| | RTH_Sorenson | Sørenson coefficient |
| | RTH_PS | Proportional similarity |
| | RTH_Euclidean | Euclidean distance |
| | RTH_Morisita | Morisita index |
| | RTH_Horn | Horn's index |
| | RTH_BrayCurtis | Bray-Curtis dissimilarity |
| | RTH_Pinkham | Pinkham and Pearson's index |

**Table 65.** Structure of the spreadsheet or data table used to store diversity, evenness, and dominance indices (RTH_Diversity, Appendix II).

| Column name | Data type | Example | Comment |
|---|---|---|---|
| SUID | Text | ALBE | Study Unit identifier |
| STAID | Text | 03015795 | Station identifier |
| Reach | Text | A | Sampling reach |
| CollectionDate | Date | 6/27/1996 | Collection date |
| SampleID | Long | 7650 | Sample identifier |
| SMCOD | Text | ALBE0696IRM0001 | Sample code |
| Margalef | Double | 18.79586 | Margalef's diversity |
| Menhinick | Double | 1.289469 | Menhinick's diversity |
| SimpsonDom | Double | 0.049463905 | Simpson dominance |
| SimpsonDiv | Double | 0.950536095 | Simpson diversity |
| ShanDiv | Double | 1.483501 | Shannon diversity |
| BrillDiv | Double | 1.458924 | Brillouin's diversity |
| SimpEven | Double | 0.965008 | Simpson evenness |
| BrillEven | Double | 0.818169 | Brillouin's evenness |
| ShanEven | Double | 0.818297 | Shannon evenness |

**Table 66.** Structure of the spreadsheet or data table used to output all similarity indices to a single spreadsheet or data table (RTH_Similarity, Appendix II).

| Column name | Data type | Example | Comment |
|---|---|---|---|
| SUID | Text | NOAMBIG | Study Unit identifier |
| STAID | Text | 03015795 | Station identifier |
| Reach | Text | A | Sampling reach |
| CollectionDate | Date | 6/27/1996 | Collection date |
| SMCOD | Text | ALBE0696IRM0001 | Sample code for Samp1 |
| Samp1 | Long | 7650 | Sample identifier for sample 1 |
| Samp2 | Long | 7658 | Sample identifier for sample 2 |
| Jaccard | Double | 0.261364 | Jaccard's coefficient |
| Sorenson | Double | 0.414414 | Sørenson's coefficient |
| PS | Double | 21.17604 | Proportional similarity |
| Euclidean | Double | 0.90275 | Euclidean distance |
| Morisita | Double | 0.24583 | Morisita's index |
| Horn | Double | 0.317631 | Horn's index |
| BrayCurtis | Double | 0.78824 | Bray-Curtis dissimilarity |
| Pinkham | Double | 0.107478 | Pinkham and Pearson's index |

**Table 67.** Structure of the spreadsheet or data table used to output each similarity index to a single spreadsheet or data table (RTH_BrayCurtis, Appendix II).

| Column name | Data type | Example | Comment |
|---|---|---|---|
| SUID | Text | ACAD | Study Unit identifier |
| STAID | Text | 07375050 | Station identifier |
| Reach | Text | A | Sampling reach |
| CollectionDate | Date/time | 1/20/1999 | Collection date |
| SMCOD | Text | ACAD0199IRM0210 | Sample code |
| SampleID | Long | 58616 | Sample identifier |
| Index | Text | BrayCurtis | Similarity index |
| 58616 | Double | 1 | Sample 58616 |
| 16429 | Double | 0.056455129 | Sample 16429 |
| 58636 | Double | 0.083951104 | Sample 58636 |

## Resetting or Exiting the Module

The IDAS program indicates that the module has finished calculating and saving diversity and similarity indices by displaying the word **FINISHED** in the right-hand panel of the **status bar** (fig. 54). The user may reset the module and process another dataset by selecting File/Close from the menu bar. Clicking on Exit on the menu bar will shut down the module and return the user to the main IDAS window (fig. 1). The user also can exit this module by clicking on the 'x' in the upper right-hand corner of the module window.

# Data Export Module

The Data Export module uses spreadsheets or tables produced by the Data Preparation module (table 3) to produce ASCII text files that can be imported into spreadsheets (for example, Excel®), databases (for example, Access®), statistical packages (for example, SYSTAT, SPLUS, SAS, CANOCO, TWINSPAN, Primer-E), and graphics packages (for example, CorelDraw). This module also produces a site (columns) by taxa (rows) table that can be used to publish invertebrate community data in reports.

The Data Export module is activated by clicking on the Data Export button of the main IDAS window (fig. 1), which brings up the Format data for export window (fig. 55). The Data Export module uses standard menu items for opening and closing files (Files), viewing data files (View), executing selected processing options (Run), modifying program options (Options), exiting the module (Exit), and displaying information about the module (About). The bottom of the module has a standard five-panel **status bar** that displays (from left to right) the name of the source file, the source-file type (Excel or Access), the name of the spreadsheet or data table that is

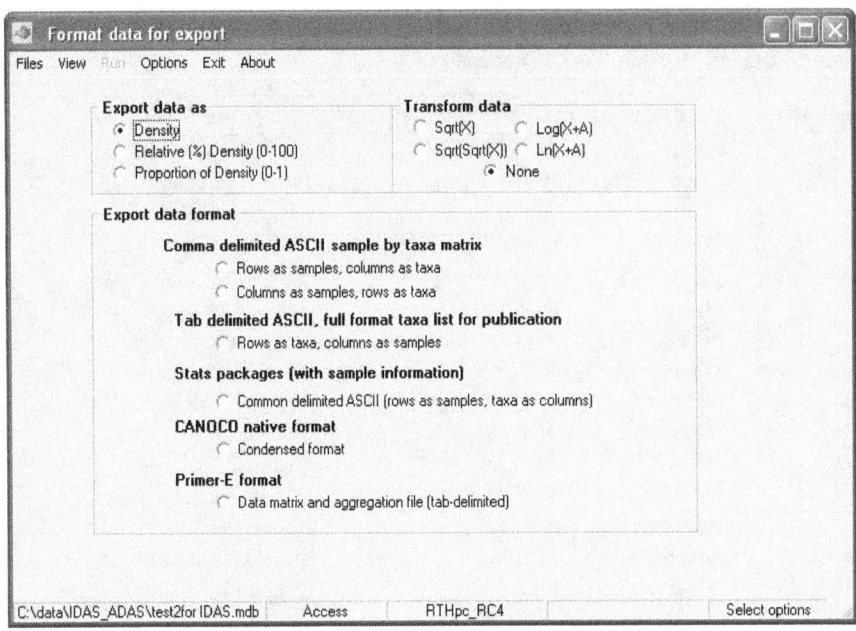

**Figure 55.** Main window of the Export Data module.

[The frames **Export data as**, **Transform data**, and **Export data format** appear after the user successfully opens a spreadsheet or data table using the Files/Open menu items]

the source of the data, the name of the spreadsheet or data table that will store the processed data, and processing status messages.

Data can be exported as abundance or density or as the proportion (0–1) or percentage (0–100) of total abundance or density in each sample. This module also allows the user to transform the data ($\log_{10}$, ln, square root, fourth root) before exporting the data in one of five formats. The export examples listed below were exported from the file ALBE_Inverts.xls:

1. **Comma-delimited ASCII sample by taxa matrix**: ASCII text files that store abundance data using abbreviations for taxa and samples and produce a second (key) file containing information on the sample and taxa abbreviations.

   a. **Rows as samples, columns as taxa**: Export_Ex1a.txt, Export_Ex1a_Key.txt

   b. **Columns as samples, rows as taxa**: Export_Ex1b.txt, Export_Ex1b_Key.txt

2. **Tab-delimited ASCII, full format taxa list for publication**: ASCII text files with the taxa in rows and samples in columns. This format includes multiple rows of sampler information and columns containing the taxonomic hierarchy: Export_Ex2.txt

3. **Stats package (with sample information)**, comma-delimited ASCII text files that store abundance data with sample information in rows and taxa (abbreviations) in columns, includes a "key" file listing the taxonomic information associated with the taxa

abbreviations used in the abundance file: Export_Ex3.txt, Export_Ex3_key.txt

4. **CANOCO native format**: ASCII text files in CANOCO-condensed format, includes a data file and a file identifying the taxa and samples: Export_Ex4.cnd, Export_Ex4_key.cnd

5. **Primer-E format**: ASCII text files that include an abundance file with samples in columns and taxa in rows (Export_Ex5.txt) and an aggregation file containing the taxonomic hierarchy for each taxon (Export_Ex5_agg.txt).

## Processing Options

Processing begins by clicking on the Files/Open menu item and selecting an Excel spreadsheet or Access table containing invertebrate abundance or density data. Once a spreadsheet or data table has been successfully opened, the **Export data as**, **Transform data**, and **Export data format** frames will appear (fig. 55).

The **Export data** as frame allows the user to convert abundance or density data to percentages (0–100%) or proportions (0–1). Percentages and proportions are calculated by dividing the abundance of a taxon by the total abundance

**TIP: Do not convert qualitative data to percentages or proportions.**

in the sample. The IDAS program does not differentiate between qualitative and quantitative samples when calculating percentages or proportions in the Export Data module. If the user elects to calculate percentages or proportions on a dataset containing qualitative samples, the resulting values will be 1/N for all taxa in a qualitative sample, where N is the total number of taxa in the sample. For this reason, it is recommended that percentages and proportions not be calculated for datasets that contain qualitative samples.

The Transform data option allows the user to apply four transformations to the data or, by selecting the option None, to keep the data in the form produced by the Data Preparation module (abundance or density). The transformations (square root, fourth root, $\log_{10} (X+A)$, or Ln $(X+A)$) are applied to the data after the options selected in Export data as are implemented. Consequently, it is possible to convert data to the square root or log of the proportions or percentages of the abundance or density data. The log transformations can add a constant (A) to the data value (X) to avoid generating an error if the data value is zero. The option None is the default for the IDAS program.

The Export data format frame provides five options for formatting the output of the abundance data produced by the Data Preparation module. Files may be exported as comma-delimited ASCII, tab-delimited ASCII, comma-delimited ASCII format for use with statistical packages, as CANOCO native (condensed) format, or Primer_E format. The comma-delimited ASCII formats are compatible with many common software packages, including SYSTAT, SPLUS, Access, and Excel. Data can be exported with rows as samples and columns as taxa (Export_Ex1a.txt) or with rows as taxa and columns as samples (Export_Ex1b.txt), depending on the needs of the user and the characteristics of the software. Some software programs (for example, Excel) have limitations on the number of rows and (or) columns that can be imported. Because expressing taxa as rows or columns affects the number of rows and columns in the resulting file, it also may affect whether the software program can import the resulting file. For example, Export_Ex1a.txt and Export_Ex1b.txt contain exactly the same data. However, Export_Ex1a.txt cannot be imported into Excel 2003 because the taxa are stored in 339 columns, which exceeds the number of columns (256) that this version of Excel can import. Appendix Export_Ex1b.txt has samples stored in 67 columns, so this table can be imported into Excel even though it contains just as many cells as Export_Ex1a.txt. The maximum number of rows in versions of Excel 2003 is about 65,198. Excel 2007 allows for larger datasets, approximately 1,048,576 rows and 16,384 columns.

Comma-delimited and CANOCO native formats use simple abbreviations to represent the names of taxa (for example, TX1, TX2, TX3) and samples (for example, S1, S2, S3). Abbreviations are used because some software programs (for example, CANOCO and SAS) require variable names to be eight characters or less. When the IDAS program uses abbreviations, it produces a "key" file that contains a list of

abbreviations, the information represented by the abbreviations, the text file to which the abbreviations apply, the date and time that the text files were created, and the name and path of the Excel or Access file that supplied the data. "Key" files are named by adding the suffix "_key" to the name used to store the exported density data. For example, if the user elects to store data for the NOAMBIG Study Unit in a file named "NOAMBIG.txt," the IDAS program provides "RTH_key.txt" as the default name for the "key" file.

> TIP: The tab-delimited, full format option is very useful for examining community data or generating reports for publication.

The user has the option of changing the default name provided by IDAS, but this will make it more difficult to associate the exported data with the list of abbreviations. Even if the user changes the name of the "key" file, the abbreviations still can be associated with the exported file because the file that contains the explanation of the abbreviations still contains the name of the text file to which the abbreviations apply and the name of the Excel or Access file from which the data were originally derived (Export_Ex1a_key.txt).

The Tab-delimited ASCII, full format option is provided to help with the process of creating data tables for reports or other purposes in which abbreviations are not appropriate. This option exports data with taxa as rows and samples as columns (Export_Ex2.txt). The rows of taxa data are sorted in phylogenetic order (SortCode) and contain all taxa levels from phylum to BU_ID as well as lifestage and SortCode data. The columns of data are identified as SUID, STAID, Reach, CollectionDate, SampleID, and SMCOD. The column header also indicates what the data represent: density (Density), proportions of density (Prop_Density), or percentages of density (%_Density). This format, when imported into Excel, provides an easy way to view and compare datasets.

The Stats packages option is a comma-delimited ASCII format with rows as samples and taxa as columns (Export_Ex3.txt). This option produces a data file and a "key" file (Export_Ex3_key.txt) that explains the abbreviations used in the data file. The Stats packages option differs from the comma-delimited option in that it includes site information as part of the file. This site information can be used by statistical packages to subset or group the data for analysis. The "key" file produced by the Stats packages option contains the names associated with the taxa abbreviations and repeats the site information contained in the abundance file. "Key" files are named by adding the suffix "_key" to the name used to store the exported density data.

The CANOCO native format option produces a file in Cornell Ecology Package (CEP)-condensed format (Export_Ex4.cnd). The CEP-condensed format produces a very compact data file that is consistent with the format of the 80-column punch cards that were used in the 1970s when the CEP was developed. Despite the antiquity of this format, it still is used by several important software packages that are frequently used for community ordinations. The files produced by this option will work with the DOS® and MS Windows® versions of CANOCO, MVSP, and PC-ORD, and the DOS version of CEP (including TWINSPAN), MVSP, and PC_ORD. CANOCO native format files are stored with the extension *.cnd to distinguish them from the other comma- and tab-delimited text files (*. txt) created by the Data Export module. PC-ORD does not recognize the file extension ".cnd;" therefore, the extension must be changed to "*.txt" before the CANOCO native format files can be used with PC-ORD.

The CANOCO-condensed file format begins with a header line (maximum of 80 characters) that identifies the file contents. The IDAS program prompts the user to enter a title line or to accept the default header line provided by IDAS (fig. 56). The CANOCO native format option uses abbreviations for taxa and sample names because the CANOCO program has an eight-character restriction on variable names. As in other cases where IDAS uses abbreviations, this option produces a "key" file that contains a list of the abbreviations, the information represented by the abbreviations, the file associated with these abbreviations, and the name and path of the Excel or Access file that provided the original data. The "key" file is stored with the same *.cnd extension that is used to store the data. Assuming that the user chooses to store the data in a file named "RTH.cnd," the IDAS program will use "RTH_key.cnd" as the default name for the "key" file.

The Primer-E format option creates two tab-delimited text files: a file of abundance data (*.txt) and an aggregation file (*_agg.txt) that contains the taxonomic hierarchy, for example Export_Ex5.txt and Export_Ex5_agg.txt, respectively. STAID or SampleID can be used to identify columns of data in the Primer output by setting the default value for ID samples in Primer output using in the Options menu (fig. 2) to Site ID (STAID). Selecting STAID as the default option facilitates relating site-specific environmental data (for example, land cover), which are typically stored under the station identifier (STAID), to the biological data, which are typically stored under the sample identifier (SampleID). The STAID can be substituted for the SampleID only when the datasets contain one sample per site. If the user elects to identify samples by STAID, the IDAS program checks to make sure that the STAID provides a unique sample identifier before substituting STAID for SampleID. If the STAID is

not a unique sample identifier, IDAS will warn the user and proceed to use SampleIDs to identify samples. Full taxa names (BU_IDs) are used to identify taxa (rows), and the export files include a title row as the first line of data (fig. 56). Other sample identification information (SUID, SMCOD, Reach, CollectionDates, and so forth) is included as factors at the bottom of the abundance matrix. The aggregation matrix (taxonomic hierarchy) includes data from all taxonomic levels, BU_ID to phylum. The Primer export data can be imported into Primer 6 using the File and Open menus in Primer and the settings listed in table 68 for abundance and aggregation files.

**Table 68.** Settings to use when opening abundance and aggregation files in Primer 6.

[NA, not applicable]

| Setting | Abundance | Aggregation |
|---|---|---|
| Data type | Sample data | Aggregation data |
| Title | Yes | Yes |
| Shape | Rectangular | NA |
| Orientation | Samples as columns | NA |
| Data type | Abundance | NA |
| Blank= | Missing value | NA |
| Text delimiters | Tab | Tab |
| Text encoding | ASCII | ASCII |

## Duplicate Sort Codes

Sort codes (SortCode) in Bio-TDB are not static; they change over time as new taxa are added to the master taxa list maintained by the NWQL BG. Consequently, it is possible to introduce duplicate sort codes into IDAS datasets by combining data (for example, data from multiple Study Units) that were downloaded from Bio-TDB at different times. The IDAS program addresses this problem by checking files for duplicate sort codes every time it opens a data file. If the IDAS program finds duplicate sort codes, it will warn the user and provide a list of taxa that have duplicate codes (fig. 57). Duplicate sort codes do not affect the calculations performed by the IDAS program, they only affect how the data are sorted in the data tables and spreadsheets created by IDAS. Duplicate sort codes are a problem in the Data Export module because they can produce multiple lines of data for a single taxon and corrupt the taxonomic order in the data file. If the Data Export module detects duplicate sort codes, it will alert the user and list the taxa with multiple sort codes (fig. 57). The user can exit the module (Exit button) and manually check SortCodes in the data or continue to export the data (Continue button) with the duplicate sort codes. While it is possible to produce datasets with multiple sort codes, the algorithms built into the IDAS program should ensure that this will be an infrequent occurrence.

**Figure 56.** The IDAS program prompts the user to supply a header line for files stored in CANOCO native format files.

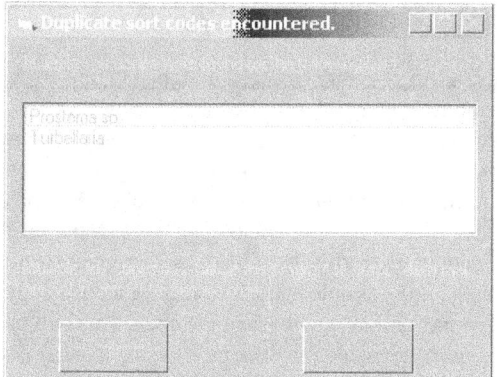

**Figure 57.** A message box alerts the user if duplicate sort codes are present in the data.

## Resetting or Exiting the Module

The IDAS program indicates that the module has finished exporting data by displaying the word **FINISHED** in the right-hand panel of the **status bar**. The user may reset the module and process another dataset by selecting File/Close from the menu bar. Clicking Exit on the menu bar will close this module and return the user to the main IDAS window (fig. 1). The user also may exit this module by clicking on the "x" in the upper right-hand corner of the module window.

# Troubleshooting

The IDAS program has been tested on a variety of computers running different versions of Windows (NT,® 2000,® XP,® Vista®) and MS Office® (2003, 2007) under a variety of computing environments typically found in USGS and home offices. Numerous data-checking and error-trapping subroutines have been built into the program in an effort to detect problems before they affect the program or after the program detects them. This section describes the various error-handling procedures built into the IDAS program and what to do when the program abnormally terminates.

## Types of Errors

The IDAS program has been designed to handle various situations where errors may arise. Errors can be divided into three categories: anticipated, unanticipated but trappable, and unanticipated and untrappable errors. Anticipated errors are those that normally arise from inappropriate actions by the user, including selecting files with the wrong formats, selecting combinations of options that generate no data, or saving data

to spreadsheets or data tables that already exist. The IDAS program includes subroutines that detect these errors, alerts the user that an error has occurred, and provides an opportunity to correct the error (fig. 58).

Unanticipated but trappable errors can occur when a file selection, query, or calculation generates a system error. For example, division by zero generates an error, as does trying to open a file that is already in use by another program. The IDAS program traps these errors and reports them back to the user (fig. 59). These errors generate an error message window that includes a header that tells where the error occurred in the program (for example, Calculate Diversities and Similarities module), the error number used by Visual Basic (#3075), and a brief description of the error (a syntax error in a query). When a trappable error is encountered, it is important for the user to copy down all the information included in the error message and report it to the author as soon as possible. The user

**Figure 58.** Error message generated by an anticipated and trappable error.

[This type of error message describes the error and lists the variable(s) that produced the error so the user can correct the problem that is causing the error. In this example, the program detected missing values in the CollectionDate column]

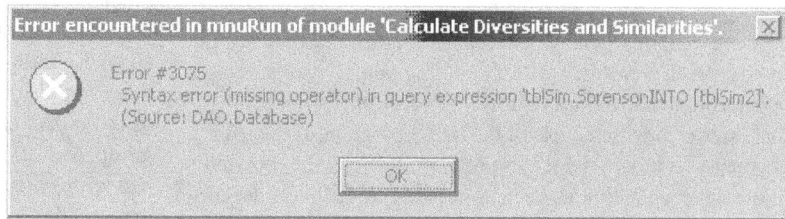

**Figure 59.** Error message generated by an unanticipated but trappable error.

[In this example, a missing data value caused an error in a database query]

should also include a complete description of what was being done when the error occurred, the data file that caused the error, and the options that were selected (see Reporting Program Bugs).

Unanticipated and untrappable errors usually occur when a conflict arises between the IDAS program and another program. For example, opening Excel® and attempting to read a data file that is being processed by the IDAS program will generate errors in Excel that affect the IDAS program but cannot be detected by the IDAS program. This type of error can cause the IDAS program to terminate abnormally (see Abnormal Termination of IDAS). These types of errors should be extremely rare. If they occur, however, the user should notify the author as soon as possible and provide a description of what happened, what programs were being used, the IDAS options that were selected, and the data file that was being used. This type of error can also occur when system resources (for example, memory and hard disk space) are low. Keeping the size of the Excel workbooks small and keeping as few applications open as possible will help reduce errors related to resource problems.

## Error Messages

The IDAS program can generate a large number of error messages of the type described in figure 58. The content of these error messages varies depending on the circumstances that led to the error. Error messages are intended to be descriptive: that is, they should describe what caused the error, where it occurred in the program, and what actions the users should take to deal with the error. They are also designed to allow the user to resolve the error without causing an unexpected termination of the program. Many of the error messages generated by the IDAS program address problems with the format or content of the data files. Consequently, IDAS contains extensive data checking subroutines that can discover problems with the data and recommend solutions before these problems affect computations. The list of possible error messages is too extensive to review in this manual; however, they should provide the user with enough information to resolve the error either manually or through tools provided as part of IDAS.

## Reporting Program Bugs

A program bug may be a calculation that does not appear to be correct, a problem with formatting output, user messages or interfaces that are not intuitive, or some other problem that interferes with the use of the IDAS program. If the user encounters a bug in the IDAS program; it should be reported to the program author as soon as possible. A complete description of the problem, any error messages that were displayed, a copy of the data file that was being processed, and a list of options that were selected should be provided.

## Abnormal Termination of IDAS

The IDAS program has been designed to trap errors and exit the module where the error occurred. However, it is not possible to anticipate all potential sources of error. Therefore, on rare occasions, the IDAS program may terminate abnormally. Abnormal termination occurs when the IDAS program terminates without the user clicking on the Exit button from the opening window (fig. 1). If the IDAS program terminates abnormally, it is possible that it will leave a hidden copy of Excel running and the source data file open. This will prevent the user from accessing the data file by using another copy of Excel because the data are already in use by the first copy of Excel. If the IDAS program terminates abnormally, the MS Task Manager® can be used to see if one or more copies of Excel (EXCEL.EXE) or IDAS (IDAS.EXE) are still running. First, the user should make sure that visible copies of Excel are closed and then open Task Manager (ctrl+alt+del, select Task Manager) and select the "Processes" tab. Click on the "Image name" column header to arrange the processes that are running in alphabetical order. Click on entries for EXCEL.EXE or IDAS.EXE and click on the "End Process" button to terminate the process. Close Task Manager and restart IDAS. Alternatively, open processes can be removed by restarting the computer. Abnormal termination also will leave a temporary Access file on the user's computer. This file (T%%%Temp.mdb) will be located in the directory that was the source of the invertebrate abundance data spreadsheet or data table. The temporary Access file is used by IDAS to temporarily store data, and calculations and will be automatically deleted the next time IDAS processes a data file from this directory, so the user does not need to remove this file manually.

## Summary

The Invertebrate Data Analysis System (IDAS) program provides an accurate, consistent, efficient, and user-friendly mechanism for analyzing invertebrate data and for exporting data to other computer programs. It provides standardized tools to capture, edit, analyze, and export data downloaded from the NAWQA Program's Biological Transactional Database (Bio-TDB) or imported from non-NAWQA sources. The format of data exported from Bio-TDB is optimized for efficient storage and must be manipulated before it can be imported into other data analysis software packages or into tables for publication. The IDAS program provides the tools to accomplish these manipulations and to access a variety of ecological data needed to calculate tolerance, functional group, and behavioral metrics. This software package reads and writes data files in the Microsoft Excel® or Access® format exported by Bio-TDB or converted to this format using the IDAS data import functions. IDAS version 5 can read and write Excel and Access files in either Microsoft Office® 2003 or 2007 formats (*.xls, *.xlsx, *.mdb, *.accdb). All this can be done in the five modules that make up the IDAS program—Edit Data, Data Preparation, Calculate Community Metrics, Calculate Diversities and Similarities, and Data Export modules.

The Edit Data module allows the user to subset data into new spreadsheets or tables on the basis of sample (combinations of SUID, STAID, reach, collection date, SampleID, SMCOD, or sample type) or taxonomic information (for example, major orders of insects, major families of Diptera). Spreadsheets or tables with similar formats can be combined into new spreadsheets or tables, and spreadsheets or tables of any type can be deleted. The summarize taxa function counts the number of sites where each taxon is found, the number of samples containing each taxon, the abundance of the taxon, and basic statistics (mean, maximum, minimum, and standard deviation) for taxon abundance across all sites and samples and across only those sites or samples in which the taxon occurs. The summarize taxa function also identifies ambiguous taxa, counts the number of ambiguous children associated with each ambiguous taxon, sums the abundances of the ambiguous children, and calculates the percentage of total abundance composed of these children. Ambiguous and provisional/conditional taxa can be resolved in the Edit data module as well as in the Data Preparation module. Files of taxa attributes (tolerances, functional groups, and behavioral traits) can be created and maintained in the Edit data module, which also provides tools for importing tolerance and functional group data from non-NAWQA sources. Data from sources other than Bio-TDB can be imported using user-defined data conversion formats. Data in matrix format (taxa-by-sample, or sample-by-taxa) can be converted into stacked column format and imported into IDAS. Random subsamples of data can be extracted from datasets and analyzed. The Edit Data module is the only IDAS module that can process data files downloaded from Bio-TDB (raw format), produced by the Data Preparation module (processed format), or imported from a non-NAWQA source.

The Data Preparation module prepares data for analysis. This module processes data exported from Bio-TDB or converted to Bio-TDB format and produces a new "processed" format that can be read by the other IDAS modules. The Data Preparation module allows the user to select the type of samples to process (RTH, DTH, QMH, and (or) QUAL), calculates densities (number/m$^2$), deletes taxa based on laboratory processing notes, deletes or retains lifestage information, sets a lowest taxonomic level for analysis, deletes rare taxa based on number of sites where the taxon occurs and (or) the percentage abundance of the taxon in each sample, and resolves taxonomic ambiguities above a user-specified taxonomic level and (or) by using one of four methods based on separate analyses for each sample or an analysis of the combined dataset. Ambiguous taxa need to be resolved prior to calculating community metrics, because the redundant taxonomic information represented by ambiguous taxa can adversely affect the calculation of richness and abundance metrics and indices. This Data Preparation module can also

produce a synthetic sample (QUAL) that represents a list of taxa found in QMH, RTH, and (or) DTH samples collected at a site within a user-specified number of days (± 7 days) from the QMH collection date. Processed data are saved to a new spreadsheet or data table in the database or workbook that provided the Bio-TDB data. This module is self-documenting and stores information on data sources, processing options, and destination files in the "Notes" spreadsheet of the attributes file (Attribute_BEHAV_v5a.xls) or in the "_Options" data table or spreadsheet associated with the data file containing the abundance information.

The Calculate Community Metrics module uses data produced by the Data Preparation module. The Calculate Community Metrics module calculates 25 metrics based on taxa richness, 24 metrics based on percentage of taxa richness, 25 metrics based on abundance, 24 metrics based on percentage of abundance, 5 dominance metrics, 32 functional group metrics (8 richness, 8 percentage richness, 8 abundance, and 8 percentage abundance, plus the percentage of taxa richness and abundance that were assigned to a functional group), 28 behavioral metrics (7 richness, 7 percentage richness, 7 abundance, 7 percentage abundance, plus the percentage of taxa richness and abundance that were assigned a behavioral trait), and 16 tolerance metrics (mean tolerance; abundance-weighted tolerance, Hilsenhoff Biotic Index; 12 tolerance classes based on richness, percentage richness, abundance, and percentage abundance of taxa defined as intolerant, moderately tolerant, or tolerant; plus the percentage of taxa richness and abundance that were assigned a tolerance value). This module uses an ecological attribute file (Attributes_BEHAV_v5a.xls) that contains data on functional feeding groups, behavioral traits, regional ecological tolerances, and a national tolerance value that is the mean of the regional values. The IDAS program allows the user to calculate tolerance values based on national or regional tolerance values. The attribute file also contains a list of equivalent taxa that allows the user to substitute functional feeding group or tolerance values reported at higher taxonomic levels (for example, genus instead of species) when a value is not available at a lower level (for example, substitute data for *Hydropsyche* when data are unavailable for *Hydropsyche sparna*). The metrics produced by this module are stored as separate spreadsheets or data tables in the workbook or database that provided the abundance data. This module is self-documenting and stores information on data sources, processing options, and destination files in the "_Options" data table or spreadsheet.

The Calculate Diversities and Similarities module uses data files produced by the Data Preparation module. This module can calculate five diversity indices (Margalef, Menhinick, Simpson, Brillouin, and Shannon), a dominance index (Simpson), three measures of evenness (Simpson, Brillouin, and Shannon), two qualitative similarity indices (Jaccard and Sørensen), four quantitative similarity indices (Proportional, Morisita, Horn, and Pinkham and Pearson), and two quantitative dissimilarity indices (Euclidean distance and Bray-Curtis). Similarity and dissimilarity indices can be saved in one of two ways: (1) in a single data table or spreadsheet with rows corresponding to sample pairs and columns to indices or (2) in a data table for each selected index with rows and columns corresponding to sample pairs. The latter format produces a symmetric matrix with the lower half populated with the similarity indices. Results from this module are stored as separate spreadsheets or tables in the workbook or database that provided the invertebrate data. This module is self-documenting and stores information on data sources, processing options, and destination files in the "_Options" data table or spreadsheet.

The Data Export module uses data files produced by the Data Preparation module. The Data Export module produces ASCII text files in comma-delimited, tab-delimited, CANOCO-condensed, or Primer-E formats that can be imported into other statistical, graphics, or word-processing programs. Data can be exported in their original format (that is, abundance or density) or converted to percentage (0–100 percent) or proportion (0–1) of sample abundance or density and then exported. Data also can be transformed based on $\log_{10}$, ln, square root, or fourth root. Comma-delimited and CANOCO-condensed format files produce two output files: (1) a data file that contains the data (that is, abundance, density, or presence) with abbreviations (eight-characters or less) for taxa and sample names and (2) a "key" file that contains a list of the abbreviations and the full sample information and taxa names. A header in the "key" file stores the name of the data file produced by the Data Export module, the time and date that the file was created, and the name and path of the spreadsheet or data table from which the data originated. This ensures that the source and data export files (data and key) can be easily associated and archived. Data also can be exported as full-format tab-delimited ASCII files with rows containing taxa and columns corresponding to samples. This format is intended to provide a table that can be imported into a word processor for publication. This format contains the full taxonomic hierarchy reported for each taxon, the taxonomic sort code, and the lifestage. Each column contains sample identification information (SUID, STAID, reach, collection date, SampleID, SMCOD) and the units of measurement (abundance or density). The rows of taxonomic data are sorted in phylogenetic order. The stats package option exports two comma-delimited files similar to the other comma-delimited files except that the abundance file contains sample information that is repeated in the "key" file along with information of taxa abbreviations. The Primer-E export format creates two files, a data file containing the abundance and sample information and an aggregation file containing the taxonomic hierarchy data. These can be exported with the STAID (station identifier) or the SampleID (sample identifier) as the sample identifier. Using the STAID can simplify matching environmental data with biological data in Primer-E. This module is self-documenting and stores information on data sources, processing options, and destination files in the "Notes" data table or spreadsheet.

# References Cited

Barbour, M.T., Gerritsen, J., Snyder, B.D., and Stribling, J.B., 1999, Rapid bioassessment protocols for use in streams and wadeable rivers—Periphyton, benthic macroinvertebrates, and fish (2d ed.): U.S. Environmental Protection Agency, Office of Water, EPA 841-B-99-002.

Brower, J.E., and Zar, J.H., 1984, Field and laboratory methods for general ecology (2d ed.): Bubuque, Iowa, William C. Brown Publishers, p. 226.

Cao, Y., Larsen, D.P., and Thorne, R. St.-J., 2001, Rare species in multivariate analysis for bioassessment—Some considerations: Journal of the North American Benthological Society, v. 20, no. 1, p. 144–153.

Cao Y., and Williams, D.D., 1999, Rare species are important for bioassessment—Reply to Marchant's comments: Limnology and Oceanography, v. 44, p. 1841–1842.

Cao Y., Williams, D.D., and Williams, N.E., 1998, How important are rare species in aquatic community ecology and bioassessment?: Limnology and Oceanography, v. 43, p. 1403–1409.

Cuffney, T.F., Bilger, M.D., and Haigler, A.M., 2007, Ambiguous taxa—Effects on the characterization and interpretation of invertebrate assemblages: Journal of the North American Benthological Society, v. 26, p. 286–307.

Cuffney, T.F., Gurtz, M.E., and Meador, M.R., 1993, Methods for collecting benthic invertebrate samples as part of the National Water-Quality Assessment Program: U.S. Geological Survey Open-File Report 93–406, 66 p.

Faith, D.P., and Norris, R.H., 1989, Correlation of environmental variables with patterns of distribution and abundance of common and rare freshwater macroinvertebrates: Biological Conservation, v. 50, p. 77–98.

Gilliom, R.J., Alley, W.M., and Gurtz, M.E., 1995, Design of the National Water-Quality Assessment Program—Occurrence and distribution of water-quality conditions: U.S. Geological Survey Circular 1112, 33 p.

Goff, F.G., 1975, Comparison of species ordinations resulting from alternative indices of interspecific association and different numbers of included species: Vegetatio, v. 31, p. 1–14.

Hirsch, R.M., Alley, W.M., and Wilber, W.G., 1988, Concepts for a National Water-Quality Assessment Program: U.S. Geological Survey Circular 1021, 42 p.

Leahy, P.P., Rosenshein, J.S., and Knopman, D.S., 1990, Implementation plan for the National Water-Quality Assessment Program: U.S. Geological Survey Open-File Report 90–174, 10 p.

Marchant, R., 1990, Robustness of classification and ordination techniques applied to macroinvertebrate communities from the La Trobe River, Victoria: Australian Journal of Marine and Freshwater Research, v. 41, p. 493–504.

Marchant, R., Hirst, A., Norris, R.H., Butcher, R., Metzeling, L., and Tiller, D., 1997, Classification and ordination of macroinvertebrate assemblages from running waters in Victoria, Australia: Journal of the North American Benthological Society, v. 16, p. 664–681.

Moulton, S.R., II, Carter, J.L., Grotheer, S.A., Cuffney, T.F., and Short, T.M., 2000, Methods for analysis by the U.S. Geological Survey National Water Quality Laboratory— Processing, taxonomy, and quality control of benthic macroinvertebrate samples: U.S. Geological Survey Open-File Report 00–212, 49 p.

Moulton, S.R., II, Kennen, J.G., Goldstein, R.M., and Hambrook, J.A., 2002, Revised protocols for sampling algal, invertebrate, and fish communities as part of the National Water-Quality Assessment Program: U.S. Geological Survey Open-File Report 02–150, 75 p.

North Carolina Department of Environment and Natural Resources, 2006, Standard operating procedures for benthic invertebrates: Division of Water Quality, accessed September 28, 2009, at *http://www.esb.enr.state.nc.us/BAUwww/benthossop.pdf*.

Washington, H.G., 1984, Diversity, biotic and similarity indices, A review with special relevance to aquatic ecosystems: Water Research, v. 18, no. 6, p. 653–694.

# Appendix I: Descriptions and Formulas Used to Calculate Diversity and Similarity Indices

## IA. Diversity and evenness indices

**Margalef's diversity:** a simple index that takes into account richness and abundance in the sample, but does not consider how the sample abundance is distributed among taxa.

$$D_q = (S{-}1)/\log_{10}(N)$$

**Menhinick's diversity:** a simple index that takes into account richness and abundance in the sample, but does not consider how the sample abundance is distributed among taxa.

$$D_b = S/\sqrt{N}$$

**Simpson's dominance:** a measure of the probability that two individuals selected randomly from a community will be the same species.

$$l_i = (\Sigma n(n{-}1))/(N(N{-}1))$$

**Simpson's diversity:** the inverse of the Simpson's dominance index. It is an expression of the number of times that two individuals can be picked at random from a community without selecting two individuals of the same species. This index is a more comprehensive measure of diversity than Margalef's and Menhinick's diversity indices because it considers richness, abundance, and how abundance is distributed among species.

$$D_s = 1{-}(\Sigma n(n{-}1))/(N(N{-}1))$$

**Brillouin's diversity:** information theory-based index that measures the "uncertainty" of a taxon selected at random from the community. High diversity is associated with high uncertainty and low diversity with low uncertainty. This index is the equivalent of the Shannon diversity index, but it is intended to be used when abundance data come from a nonrandom sample or when the collected data constitute the entire community or subcommunity.

$$H = (\log_{10}(N!){-}\Sigma\log_{10}(n!))/N$$

**Shannon's diversity:** information theory-based index that measures the "uncertainty" of a taxon selected at random from the community. High diversity is associated with high uncertainty and low diversity with low uncertainty. This index is the equivalent of the Brillouin's diversity index, but it is intended for use when the abundance data come from a random sample of the community or subcommunity.

$$H` = (N \log_{10} N - \Sigma\, n \log_{10} n)/N$$

**Simpson's evenness:** ratio of the observed Simpson diversity to the maximum possible diversity (that is, diversity when individuals are distributed as evenly as possible among the species).

$$E_s = D_s/D_{max} \quad \text{where } D_{max} = ((S-1)/S)/(N/(N-1))$$

**Brillouin's evenness:** ratio of the observed Brillouin diversity to the maximum possible diversity (that is, diversity when individuals are distributed as evenly as possible among the species). Like Brillouin's diversity index, this measure is intended to be used when the abundance data come from a nonrandom sample or when the collected data constitute the entire community or subcommunity.

$$J = H/H_{max} \quad \text{where } H_{max} = [\log_{10} N! - (S-r) \log_{10} c! - r \log_{10} (c+1)!]/N$$

**Shannon's evenness:** ratio of the observed Shannon diversity to the maximum possible diversity (that is, diversity when individuals are distributed as evenly as possible among the species). Like the Shannon diversity index, this measure is intended to be used when the abundance data come from a random sample or the community of subcommunity.

$$J` = H`/H_{max}` \quad \text{where } H_{max}` = \log_{10} S$$

**Abbreviations used in formulas:**

$S$ = number of taxa in sample
$n$ = abundance of an individual taxon
$N$ = total number of individuals in sample
$c$ = integer portion of N/S
$r$ = remainder of N/S

## IB. Similarity indices

**Jaccard coefficient:** a qualitative measure of similarity that simply expresses the percentage of species shared in common between two communities. It considers only the number of taxa in the two communities, not their abundances. Values range from 0 (no species found in both communities) to 1 (all species found in both communities). This index is useful for qualitative samples (QUAL and QMH), but measures incorporating information on the abundance and the distribution of abundance among species are more appropriate for qualitative (RTH and DTH) samples.

$$CC_j = c/(s_1+s_2-c)$$

**Sørensen coefficient:** a qualitative measure of similarity that is similar to the Jaccard coefficient in that it expresses the percentage of species shared in common between two communities. It considers only the number of taxa in the two communities, not their abundances. Values range from 0 (no species found in both communities) to 1 (all species found in both communities). This index is useful for qualitative samples (QUAL and QMH), but measures incorporating information on the abundance and the distribution of abundance among species are more appropriate for quantitative (RTH and DTH) samples.

$$CC_s = 2c/(s_1+s_2)$$

**Proportional similarity:** a quantitative measure of similarity that compares the percentage composition of two communities. This index incorporates information on richness and abundances expressed as a percentage of total abundance. Proportional similarity would be appropriate to use when the analyst wants to concentrate on similarities in the structure of the community (relative abundances of taxa) rather than on differences in abundances. This index varies from 0 (completely dissimilar communities) to 100 (identical communities).

$$PS = \Sigma(\min[p_1,p_2])$$

**Euclidean distance:** a measure of how far apart two communities are in species composition. This index incorporates information on richness, abundance, and the distribution of abundances among species. Unlike the proportional similarity index, Euclidean distance is sensitive to differences in both relative and absolute abundances between communities. This index measures how dissimilar communities are rather than how similar. This index varies from 0 (identical communities) to 1 (completely dissimilar communities).

$$I_3 = \sqrt{(\Sigma((x_i-y_i)/(x_i+y_i))^2/S)}$$

**Morisita's similarity:** a quantitative measure of similarity that is based on Simpson's dominance index. This index incorporates information on richness, abundance, and the distribution of abundances among species. In contrast to Euclidean distance, Morisita's index is affected very little by the sizes of the samples. It represents the probability that individuals randomly drawn from each of the two communities will belong to separate species, relative to the probability of randomly selecting a pair of specimens of the same species from one of the communities. This index varies from 0 (completely dissimilar communities) to approximately 1 (identical communities).

$$I_M = (2\Sigma x_i y_i)/((l_1+l_2)N_1N_2) \quad \text{where} \quad l_1 = (\Sigma x_i(x_i-1))/(N_1(N_1-1))$$
$$l_2 = (\Sigma y_i(y_i-1))/(N_2(N_2-1))$$

**Horn's similarity:** a quantitative measure of similarity that is based on information theory. This index incorporates information on richness, abundance, and the distribution of abundances among species. It is sensitive to differences in abundances between communities. The index varies from 0 (completely dissimilar communities) to 1 (identical communities).

$$R_o = (H_4\grave{} - H_3\grave{})/(H_4\grave{} - H_5\grave{}) \quad \text{where} \quad$$

$$H_5\grave{} = (N_1 H_1\grave{} + N_2 H_2\grave{})/N$$
$$H_4\grave{} = (N \log_{10} N - \Sigma x_i \log_{10} x_i - \Sigma y_i \log_{10} y_i)/N$$
$$H_3\grave{} = [N \log_{10} N - \Sigma(x_i + y_i) \log_{10} (x_i + y_i)]/N$$
$$H_2\grave{} = (N_2 \log_{10} N_2 - \Sigma y_i \log_{10} y_i)/N_2$$
$$H_1\grave{} = (N_1 \log_{10} N_1 - \Sigma x_i \log_{10} x_i)/N_1$$

**Bray-Curtis dissimilarity:** a quantitative measure of dissimilarity based on the percentage of each community. Because this index is based on percentage composition, it emphasizes differences in the structure of the communities rather than differences in the abundances. That is, community A may differ from community B only in that the abundances of species in community A are 10 times those in community B. Bray-Curtis dissimilarity would not detect these differences in abundances because the percentage compositions are the same in each community. This may be advantageous when comparing communities collected using different techniques. Values range from 0 (identical communities) to 1 (completely dissimilar communities).

$$D = 0.5(\Sigma |p_1/100 - p_2/100|)$$

**Pinkham and Pearson's similarity:** a quantitative measure of similarity that compares species compositions simultaneously. This index incorporates information on richness, abundance, and the distribution of abundances among species. It is sensitive to differences in abundances between communities and differences in relative abundances. Values range from 1 (identical communities) to 0 (completely dissimilar communities).

$$B = (1/K) \ \Sigma \ (\min(x_i, y_i)/\max(x_i, y_i))$$

**Abbreviations used in formulas:**

$c$ = the number of taxa that occur in both sample 1 and 2
$s_1$ = the number of taxa in sample 1
$s_2$ = the number of taxa in sample 2
$\min$ = minimum
$\max$ = maximum
$p_1$ = percentage abundance of taxon in sample 1
$p_2$ = percentage abundance of taxon in sample 2
$x_i$ = abundance of species "i" in sample 1
$y_i$ = abundance of species "i" in sample 2
$S$ = the number of taxa in both communities
$N_1$ = number of individuals in sample 1
$N_2$ = number of individuals in sample 2
$N$ = number of individuals in samples 1 and 2 $(N_1 + N_2)$

**Note:** The IDAS program uses $\log_{10}$ to calculate diversity indices. These indices can be converted to other bases by multiplying the IDAS index by the following factors.

$\log_e$: 2.3036
$\log_2$: 3.3219

# Appendix II: Example of Files Used and Produced by IDAS

| Files used or produced by IDAS | Contents | Description |
|---|---|---|
| **Bio-TDB data files:** | | |
| ALMN_05092008_1037_Invert_Results_Comb.xls | ALMN | Invertebrate data for the ALMN Study Unit |
| ALMN_05122008_1001_Sample_All.xls | ALMN_1 | Sample areas for the ALMN Study Unit invertebrate data |
| IDAS_20090715_1104_tcuffney.mdb | Invert_Results_Comb | Invertebrate data for the ALBE Study Unit |
| | Sample_All | Sample areas |
| | Bio_TFB_Criteria | Criteria used to extract data from Bio-TDB |
| | RTH | RTH processed through the Data Preparation module |
| | RTH_Stats | Processing statistics for creation of NoAmbig |
| | RTH_Options | Documentation of processing options applied to NoAmbig |
| | RTH_Distrib | Summarization of the distribution of taxa in NoAmbig |
| | RTH_R_Metrics | Richness metrics extracted from NoAmbig |
| | RTH_pR_Metrics | Percentage richness metrics extracted from NoAmbig |
| | RTH_A_Metrics | Abundance metrics extracted from NoAmbig |
| | RTH_pA_Metrics | Percentage abundance metrics extracted from NoAmbig |
| | RTH_DOM_Metrics | Dominance metrics |
| | RTH_DOM_Taxa | Dominant taxa |
| | RTH_TOL_Metrics | Tolerance metrics |
| | RTH_FG_Metrics | Functional group metrics |
| | RTH_Diversity | Diversity indices |
| | RTH_Similarity | Similarity indices combined in one file |
| | RTH_Pinkham | Pinkham similarity index, one index per spreadsheet |
| | QUAL | Qualitative sample formed by combining QMH and RTH samples |
| | QUAL_Options | Documentation of processing options applied to QUAL |
| | QUAL_Stats | Processing statistics for creation of QUAL |
| | QUAL_QQSMCODs | RTH and QMH SMCODs combined to form QUAL |
| IDAS_20080505_1404_tcuffney_Sample_All.xls | Sample_All | Sample areas for the ALBE Study Unit invertebrate data |
| **EMAP data files (Edit Data module, Import data/WR-EMAP format):** | | |
| WR_EMAP.xls | DATA | EMAP data |
| | EMAP | RAW data converted to Bio-TDB format |
| | EMAP_Samp_Info | Sample information extracted from RAW |
| WR_EMAP_Sample_Areas.xls | EMAP | |

| Files used or produced by IDAS | Contents | Description |
|---|---|---|
| **Attribute file supplied with IDAS v. 5.0:** | | |
| Attributes_BEHAV_v5a.xls | EQTX | Equivalent taxa data |
| | Attrib | Invertebrate attributes |
| | HIER | Taxonomic hierarchy |
| | Version | Information on the creation of the attribute file |
| | Taxa_changes | Taxonomic updates for version 5a |
| | Notes | Notes on the origin of the attribute information |
| Data_Formats.mdb | tblFormat | Import data format information |
| | tblQuant | Quantitative sample identifiers |
| | tblQual | Qualitative sample identifiers |
| | tblSampleID | Identifiers that define a SampleID |
| IDAS_Defaults.txt | Not applicable | Default values for program options |
| **Attribute file optimized for NoAmbig data in IDAS_20080505_1404_tcuffney.xls (Edit data/Maintain attribute files/Create new attribute file):** | | |
| Attributes_BEHAV_v5a_SE_TOL.xls | EQTX | Attribute information optimized for NoAmbig data |
| | Attrib | Attribute information optimized for NoAmbig data |
| | HIER | Attribute information optimized for NoAmbig data |
| | Version | Attribute information optimized for NoAmbig data |
| | Taxa_changes | Attribute information optimized for NoAmbig data |
| | Notes | Attribute information optimized for NoAmbig data |
| **Import data files (Edit Data module, Import data/User defined formats):** | | |
| Import_Format_Example_1.xls | Abund_w_SampInfo_Hier | Abundance, sample information, and taxonomic hierarchy |
| Import_Format_Example_2.xls | Abund_w_SampInfo | Abundance and sample information |
| | Hier | Taxonomic hierarchy |
| Import_Format_Example_3.xls | Abund_w_Hier | Taxonomic hierarchy and abundance information |
| | SampleInfo | Sample information |
| Import_Format_Example_4.xls | Abund | Abundance information |
| | SampleInfo | Sample information |
| | Hier | Taxonomic hierarchy |
| **Data matrix conversion files (Edit Data/Utilities/Convert matrix to columns):** | | |
| Abund_Data_TR.xls | Abund_Data_TR | Taxa (row) by sample (column) matrix |
| Full_Data_TR.xls | Full_Data_TR | Taxa (row) by sample (column) matrix with detailed sample information |
| Abund_Data_TC.xls | Abund_Data_TC | Sample (row) by taxa (column) matrix |
| Full_Data_TC.xls | Full_Data_TC | Sample (row) by taxa (column) matrix with detailed sample information |
| Sample_Info.xls | SampInfo | Sample information to support importing data in matrix form |
| | Hier | Taxonomic hierarchy to support importing data in matrix form. |

| Files used or produced by IDAS | Contents | Description |
| --- | --- | --- |
| **Data export format examples (Export Data Module):** | | |
| Export_Ex1a.txt | Not applicable | Comma-delimited ASCII, samples in rows, taxa in columns |
| Export_Ex1a_Key.txt | Not applicable | Key equating abbreviations to data |
| Export_Ex1b.txt | Not applicable | Comma-delimited ASCII, taxa in rows, samples in columns |
| Export_Ex1b_Key.txt | Not applicable | Key equating abbreviations to data |
| Export_Ex2.txt | Not applicable | Full format data with sample information and taxonomic hierarchy |
| Export_Ex3.txt | Not applicable | Export for statistical packages, samples as rows, taxa as columns |
| Export_Ex3_key.txt | Not applicable | Key equating abbreviations to data |
| Export_Ex4.cnd | Not applicable | CANOCO-condensed format, abundance data |
| Export_Ex4_key.cnd | Not applicable | Key equating abbreviations to data |
| Export_Ex5.txt | Not applicable | Primer-E data invertebrate data export format |
| Export_Ex5_agg.txt | Not applicable | Primer-E aggregation data |

# Glossary

**Abundance**    The number of organisms in a sample, either for the whole sample or for each taxon.

**Access tables**    Rows and columns of data that form the basic data storage units in Microsoft Access® database files.

**Ambiguous child**    A taxon that occurs at a lower taxonomic level within a group of ambiguous taxa. For example, in a sample that contained data for Hydropsychidae, *Hydropsyche*, and *Hydropsyche sparna*; *Hydropsyche* would be the ambiguous child of the ambiguous parent Hydropsychidae, and *Hydropsyche sparna* would be the ambiguous child of the ambiguous parents Hydropsychidae and *Hydropsyche*.

**Ambiguous parent**    A taxon within a group of ambiguous taxa that occurs at a higher taxonomic level than do other taxa within the group. For example, in a sample that contained data for Hydropsychidae, *Hydropsyche*, and *Hydropsyche sparna*; both Hydropsychidae and *Hydropsyche* would be ambiguous parents of *Hydropsyche sparna*, and Hydropsychidae would be an ambiguous parent of *Hydropsyche*.

**Ambiguous taxon**    A taxon in a dataset for which data are reported at one or more lower or higher taxonomic levels within the taxonomic hierarchy. For example, in a sample that contained data for Hydropsychidae, *Hydropsyche*, and *Hydropsyche sparna*; all three taxa would be considered ambiguous.

**AreaSampTot**    The name of the column that stores the total area sampled in square centimeters ($cm^2$) for quantitative samples (RTH and DTH).

**Benthic**    Refers to bottom; for example, benthic organisms that live on or burrow into the substrate on the bottom of a stream.

**Biological Data Analysis System (BDAS)**    A USGS software package for the analysis of NAWQA Program ecological data that was developed for use on the Data General computer system.

**Biological Transactional Database (Bio-TDB)**    The database used to enter and store biological data collected as part of the NAWQA Program.

**BU_ID**    The taxonomic name provided by the Biological Group at the USGS National Water Quality Laboratory. BU_IDs may include conditional or provisional identifications.

**CANOCO**    A commercially available multivariate statistical package for the analysis of community data.

**Child**    *See* ambiguous child.

**Collection date (CollectionDate)**     The date on which a sample was collected.

**Community metric**     A numerical summarization of the characteristics of a community.

**Component**     *See* sample component.

**Conditional identification**     An organism that has been assigned to a taxon that it closely resembles, but for which it does not fully meet the published description. These identifications are approximate rather than definitive.

**Cornel Ecology Package (CEP)**     A statistical package developed in the 1970s for the multivariate analysis of community data.

**Data table**     *See* Access tables.

**Dataset**     A group of samples that are contained within an Excel spreadsheet or Access data table.

**Density**     The number of organisms in a sample divided by the area sampled in square meters ($m^2$). Expressed as density for the whole sample, taxon, or group of taxa.

**Depositional-targeted habitat (DTH)**     A habitat within the sampling reach where fine sediments (for example, sand and silt) are deposited. A composite sample from this habitat is referred to as a "DTH sample."

**Dissimilarity index**     An index that measures how different two samples are based on the kinds of taxa present in the samples and (or) their abundances. Dissimilarity indices are related to similarity indices.

**Diversity index**     An index that reduces the structure of a community to a numeric value by mathematically describing how abundance is distributed among taxa in a sample. Diversity indices are related to dominance and evenness indices.

**Dominance index**     A numeric index that measures how strongly community structure is dominated by numerically abundant taxa. Dominance indices are related to diversity and evenness indices.

**DTH**     *See* depositional-targeted habitat

**Ecological tolerance**     A numeric value assigned to a taxon that indicates how well the taxon tolerates pollution. Low values indicate intolerant taxa that will disappear quickly from communities as water quality degrades. High values indicate tolerant taxa that will remain in the community as water quality degrades.

**Evenness index**   A numeric index that measures how uniformly (evenly) abundance is distributed among taxa in a sample. Evenness indices are related to dominance and diversity indices.

**Functional feeding group**   A group of taxa that have similar adaptations for feeding.

**Functional group**   *See* functional feeding group.

**Higher taxonomic level**   A position in the taxonomic hierarchy that is closer to the level of phylum than the level against which it is being compared. In IDAS the highest taxonomic level is phylum, the lowest level is species.

**Invertebrates**   Animals that do not have backbones, such as worms, clams, crustaceans, and insects.

**Lab notes**   Notes made by the NWQL BG during sample processing that document why organisms were not identified to taxonomic levels specified in the sample processing protocol.

**Laboratory notes**   *See* lab notes.

**Lifestage**   One of four stages (egg, larva, pupa, and adult) in the development of insects.

**Lower taxonomic level**   A position in the taxonomic hierarchy that is closer to the level of species than the level against which it is being compared. In IDAS, the highest taxonomic level is phylum, and the lowest level is species.

**Lowest taxonomic level**   The lowest level of the taxonomic hierarchy that will be used for an analysis. Data in levels below this level will be aggregated up to this taxonomic level. In IDAS, the highest taxonomic level is phylum, and the lowest level is species.

**Matrix format**   A data format in which the data are arranged with the samples as rows and taxa as columns or taxa as rows and samples as columns.

**Metrics**   *See* community metrics.

**Module**   A set of related analyses in IDAS.

**MVSP**   A commercially available multivariate statistical package for the analysis of community data.

**Notes**   *See* lab notes.

**Parent**   *See* ambiguous parent.

**PC-ORD**   A commercially available multivariate statistical package for the analysis of community data.

**Phylogenetic order**   The taxonomic hierarchy arranged along inferred lines of descent based on paleontological, morphological, or other evidence.

**Primer E**   A commercially available multivariate statistical package for the analysis of community data.

**Proportion**   The number or density of organisms of a particular taxon present in a sample divided by the total abundance or density in that sample. Proportions vary between 0 and 1.

**Proportional abundance**   The number of organisms of a particular taxon present in a sample divided by the total number of organisms in that sample. Proportional abundance varies between 0 and 1.

**Proportional density**   The density of organisms of a particular taxon present in a sample divided by the total density of organisms in that sample. Proportional density varies between 0 and 1.

**Provisional identification**   An organism that has been assigned to a provisional taxon reported in the literature, but the specific identify remains unknown (*Hydropsyche* sp. A). Also known as "operational taxonomic units" or "OTUs." These identifications are approximate rather than definitive.

**QMH**   *See* qualitative multihabitat.

**Qualitative multihabitat (QMH)**   A series of different habitats identified in a reach from which discrete collections of invertebrates are taken and later combined to form a composite sample. The composite sample is referred to as a "QMH sample."

**Qualitative sample (QUAL)**   A list of taxa found at a site that is formed by combining data from QMH samples with data from RTH and (or) DTH samples collected over a specified range of sampling dates.

**Rare taxa**   Taxa that occur at only a few sites or that contribute only a small fraction of the total abundance in a sample.

**Reach**   A length of stream (150–300 meters for wadeable streams; 300–1,000 meters for nonwadeable streams) that is chosen to represent a uniform set of physical, chemical, and biological conditions within a stream segment. Reach is the principal sampling unit for collecting physical, chemical, and biological data in the NAWQA Program.

**Relative abundance**   The number of organisms of a particular taxon present in a sample divided by the total number of organisms in that sample and multiplied by 100. Relative abundance varies between 0 and 100.

**Relative density**    The density of organisms of a particular taxon present in a sample divided by the total density of organisms in that sample and multiplied by 100. Relative density varies between 0 and 100.

**Richest-targeted habitat (RTH)**    A targeted habitat (usually a riffle or woody snag) in a reach where the taxonomically richest invertebrate community is theoretically located. Discrete collections of invertebrates are taken from this habitat and combined to form a composite sample. The composite sample is referred to as a "RTH sample."

**Richness**    *See* taxa richness.

**RTH**    *See* richest-targeted habitat

**Sample**    Operationally defined as all of the material and organisms collected during one application of the NAWQA Program sampling protocol for a particular sample type (for example, invertebrate RTH sample). A single sample may be subdivided during field processing to create multiple sample components.

**Sample component**    A subset of an invertebrate sample that is produced by processing a sample in the field. Field processing can produce up to four different sample components: large-rare, main-body, split, and elutriate.

**Sample identification code (SMCOD)**    A 16-character alphanumeric code that uniquely identifies each sample component.

**Sample medium**    The type of biological community being sampled (algae, invertebrates, or fish).

**SampleMediumCode**    The name of the column that holds information on the sample medium.

**Sample number**    A four-digit number that Bio-TDB uses to uniquely identify each sample component.

**Sample type**    A certain type of invertebrate sample collected in a reach from either a single targeted habitat (RTH or DTH) or multiple habitats (QMH).

**SampleType**    The name of the column that stores sample type information.

**SAS**    A commercially available statistics package.

**Similarity index**    An index that measures how alike two samples are based on the kinds of taxa present in the samples and (or) their abundances.

**Sort Code (SortCode)**    A number supplied by Bio-TDB that allows data to be sorted into phylogenetic order.

**SPLUS**    A commercially available statistics package.

**Spreadsheet**    A series of rows and columns that holds information in Microsoft Excel® files.

**Stacked column format**    A data format in which each row contains the data (abundance, taxonomic hierarchy, sample, and site information) for a single taxon. Taxa for each sample are stacked in consecutive rows.

**STAID**    Station identification number.

**SUID**    A four-letter abbreviation used to identify Study Units.

**SYSTAT**    A commercially available statistics package.

**Taxa**    Plural of taxon.

**Taxa richness**    The number of different taxa in a sample.

**Taxon**    A taxonomic group that is sufficiently distinct to be worthy of being distinguished by name and ranked in a definite category of the taxonomic hierarchy.

**Taxonomic hierarchy**    A hierarchic classification scheme that orders taxa into related groups based on similarities in morphological structure. The highest level of the hierarchy (phylum) has the most general characteristics and the lowest level (species) the most specific characteristics and greatest morphological similarities among organisms.

**Taxonomic level**    A grouping in the taxonomic hierarchy.

**Tolerance**    *See* ecological tolerance.

**TWINSPAN**    Two-Way Indicator Species Analysis. A computer software package for the multivariate analysis of community data that is part of the Cornel Ecology Package of software.

**Visual Basic**    A computer programming language for Microsoft Windows®.

**Workbook**    A collection of spreadsheets contained within one Microsoft Excel® file.

**Zoogeography**    The study of the geographic distribution of animals.